EDUCATION IN A COMPETITIVE AND GLOBALIZING WORLD

SCIENCE, TECHNOLOGY, ENGINEERING AND MATH (STEM) EDUCATION

ELEMENTS, CONSIDERATIONS AND FEDERAL STRATEGY

EDUCATION IN A COMPETITIVE AND GLOBALIZING WORLD

Additional books in this series can be found on Nova's website under the Series tab.

Additional E-books in this series can be found on Nova's website under the E-book tab.

EDUCATION IN A COMPETITIVE AND GLOBALIZING WORLD

SCIENCE, TECHNOLOGY, ENGINEERING AND MATH (STEM) EDUCATION

ELEMENTS, CONSIDERATIONS AND FEDERAL STRATEGY

NOA LEMOINE
EDITOR

nova publishers
New York

For permission to use material from this book please contact us:
Telephone 631-231-7269; Fax 631-231-8175
Web Site: http://www.novapublishers.com

Library of Congress Cataloging-in-Publication Data

ISBN: 978-1-62808-690-4

Published by Nova Science Publishers, Inc. † New York

CONTENTS

PREFACE

The term "STEM education" refers to teaching and learning in the fields of science, technology, engineering, and mathematics. It typically includes educational activities across all grade levels— from pre-school to post-doctorate—in both formal (e.g., classrooms) and informal (e.g., afterschool programs) settings. Federal policymakers have an active and enduring interest in STEM education and the topic is frequently raised in federal science, education, workforce, national security, and immigration policy debates. For example, more than 225 bills containing the term "science education" were introduced between the 102th and 112th congresses. The United States is widely believed to perform poorly in STEM education. However, the data paint a complicated picture. By some measures, U.S. students appear to be doing quite well. For example, overall graduate enrollments in science and engineering (S&E) grew 35% over the last decade. Further, S&E enrollments for Hispanic/Latino, American Indian/Alaska Native, and African American students (all of whom are generally underrepresented in S&E) grew by 65%, 55%, and 50%, respectively. On the other hand, concerns remain about persistent academic achievement gaps between various demographic groups, STEM teacher quality, the rankings of U.S. students on international STEM assessments, foreign student enrollments and increased education attainment in other countries, and the ability of the U.S. STEM education system to meet domestic demand for STEM labor. This book is intended to serve as a primer for outlining existing STEM education policy issues and programs. It includes assessments of the federal STEM education effort and the condition of STEM education in the United States, as well as an analysis of several of the policy issues central to the contemporary federal conversation about STEM education.

Chapter 1 – The term "STEM education" refers to teaching and learning in the fields of science, technology, engineering, and mathematics. It typically includes educational activities across all grade levels— from pre-school to post-doctorate—in both formal (e.g., classrooms) and informal (e.g., afterschool programs) settings. Federal policymakers have an active and enduring interest in STEM education and the topic is frequently raised in federal science, education, workforce, national security, and immigration policy debates. For example, more than 225 bills containing the term "science education" were introduced between the 102th and 112th congresses. The United States is widely believed to perform poorly in STEM education. However, the data paint a complicated picture. By some measures, U.S. students appear to be doing quite well. For example, overall graduate enrollments in science and engineering (S&E) grew 35% over the last decade. Further, S&E enrollments for Hispanic/Latino, American

Indian/Alaska Native, and African American students (all of whom are generally underrepresented in S&E) grew by 65%, 55%, and 50%, respectively. On the other hand, concerns remain about persistent academic achievement gaps between various demographic groups, STEM teacher quality, the rankings of U.S. students on international STEM assessments, foreign student enrollments and increased education attainment in other countries, and the ability of the U.S. STEM education system to meet domestic demand for STEM labor. Various attempts to assess the federal STEM education effort have produced different estimates of its scope and scale. Analysts have identified between 105 and 252 STEM education programs or activities at 13 to 15 federal agencies. Annual federal appropriations for STEM education are typically in the range of $2.8 billion to $3.4 billion. All published inventories identify the Department of Education, National Science Foundation, and Health and Human Services as key agencies in the federal effort. Over half of federal STEM education funding is intended to serve the needs of postsecondary schools and students; the remainder goes to efforts at the kindergarten-through-Grade 12 level. Much of the funding for post-secondary students is in the form of financial aid. Federal STEM education policy concerns center on issues that relate to STEM education as a whole—such as governance of the federal effort and broadening participation of underrepresented populations—as well as those that are specific to STEM education at the elementary, secondary, and postsecondary levels. Governance concerns focus on perceived duplication and lack of coordination in the federal effort; broadening participation concerns tend to highlight achievement gaps between various demographic groups. Analysts suggest a variety of policy proposals in elementary, secondary, and postsecondary STEM education. At the K-12 level, these include proposals to address teacher quality, accountability, and standards. At the post-secondary level, proposals center on efforts to remediate and retain students in STEM majors. This report is intended to serve as a primer for outlining existing STEM education policy issues and programs. It includes assessments of the federal STEM education effort and the condition of STEM education in the United States, as well as an analysis of several of the policy issues central to the contemporary federal conversation about STEM education. Appendix A contains frequently cited data and sources and Appendix B includes a selection of major STEM-related acts.

Chapter 2 – Report on National Science and Technology Council, on Federal Science, Technology, Engineering, and Mathematics (STEM) Education: 5-Year Strategic Plan, May 2013.

In: Science, Technology, Engineering and Math ... ISBN: 978-1-62808-690-4
Editor: Noa Lemoine

Chapter 1

SCIENCE, TECHNOLOGY, ENGINEERING, AND MATHEMATICS (STEM) EDUCATION: A PRIMER*

Heather B. Gonzalez and Jeffrey J. Kuenzi

SUMMARY

The term "STEM education" refers to teaching and learning in the fields of science, technology, engineering, and mathematics. It typically includes educational activities across all grade levels— from pre-school to post-doctorate—in both formal (e.g., classrooms) and informal (e.g., afterschool programs) settings. Federal policymakers have an active and enduring interest in STEM education and the topic is frequently raised in federal science, education, workforce, national security, and immigration policy debates. For example, more than 225 bills containing the term "science education" were introduced between the 102th and 112th congresses.

The United States is widely believed to perform poorly in STEM education. However, the data paint a complicated picture. By some measures, U.S. students appear to be doing quite well. For example, overall graduate enrollments in science and engineering (S&E) grew 35% over the last decade. Further, S&E enrollments for Hispanic/Latino, American Indian/Alaska Native, and African American students (all of whom are generally underrepresented in S&E) grew by 65%, 55%, and 50%, respectively. On the other hand, concerns remain about persistent academic achievement gaps between various demographic groups, STEM teacher quality, the rankings of U.S. students on international STEM assessments, foreign student enrollments and increased education attainment in other countries, and the ability of the U.S. STEM education system to meet domestic demand for STEM labor.

Various attempts to assess the federal STEM education effort have produced different estimates of its scope and scale. Analysts have identified between 105 and 252 STEM education programs or activities at 13 to 15 federal agencies. Annual federal appropriations for STEM education are typically in the range of $2.8 billion to $3.4 billion. All published inventories identify the Department of Education, National Science Foundation, and Health and Human Services as key agencies in the federal effort. Over half of federal STEM education funding is intended to serve the needs of postsecondary

* This is an edited, reformatted and augmented version of Congressional Research Service, Publication No. R42642, dated April 5, 2013.

schools and students; the remainder goes to efforts at the kindergarten-through-Grade 12 level. Much of the funding for post-secondary students is in the form of financial aid.

Federal STEM education policy concerns center on issues that relate to STEM education as a whole—such as governance of the federal effort and broadening participation of underrepresented populations—as well as those that are specific to STEM education at the elementary, secondary, and postsecondary levels. Governance concerns focus on perceived duplication and lack of coordination in the federal effort; broadening participation concerns tend to highlight achievement gaps between various demographic groups. Analysts suggest a variety of policy proposals in elementary, secondary, and postsecondary STEM education. At the K-12 level, these include proposals to address teacher quality, accountability, and standards. At the post-secondary level, proposals center on efforts to remediate and retain students in STEM majors.

This report is intended to serve as a primer for outlining existing STEM education policy issues and programs. It includes assessments of the federal STEM education effort and the condition of STEM education in the United States, as well as an analysis of several of the policy issues central to the contemporary federal conversation about STEM education. Appendix A contains frequently cited data and sources and Appendix B includes a selection of major STEM-related acts.

INTRODUCTION

The term "STEM education" refers to teaching and learning in the fields of science, technology, engineering, and mathematics. It typically includes educational activities across all grade levels— from pre-school to post-doctorate—in both formal (e.g., classrooms) and informal (e.g., afterschool programs) settings. Federal policymakers have an active and enduring interest in STEM education and the topic is frequently raised in federal science, education, workforce, national security, and immigration policy debates. The purpose of this report is to put various legislative and executive branch STEM education-related policy proposals into a useful context.

Although many observers cite the launch of the Soviet Union's Sputnik satellite in the 1950s as a key turning point for STEM education policy in the United States, federal interest in scientific and technological literacy writ large is longstanding and dates to at least the first Congress.[1] For example, in the first State of the Union address President George Washington called upon Congress to promote scientific knowledge for the sake of the republic and the polity, saying

> Nor am I less persuaded that you will agree with me in opinion that there is nothing which can better deserve your patronage than the promotion of science and literature. Knowledge is in every country the surest basis of public happiness. In one in which the measures of government receive their impressions so immediately from the sense of the community as in ours it is proportionably [sic] essential.[2]

More recent concerns about scientific and technological literacy in the United States focus on the relationship between STEM education and national prosperity and power. Since World War II, the United States has benefitted from economic and military advances made possible, in part, by a highly skilled STEM workforce. However, today the economic and social benefits of scientific thinking and STEM education are widely believed to have broad application for workers in both STEM and non-STEM occupations.[3] As such, many contemporary policymakers consider widespread STEM literacy, as well as specific STEM expertise, to be critical human capital competencies for a 21st century economy.[4]

The primary domestic source of STEM labor in the United States is the education system.[5] Federal legislators have paid close attention the STEM-related outputs of that system—such as the number of college graduates with degrees in STEM fields or the performance of U.S. students on international mathematics tests—and have sought to increase its functioning and capacity though federal policy and investments. For example, over 225 bills containing the term "science education" were introduced in the 20 years between the 102nd (1991-1992) and 112th (2011-2012) congresses. Agency reauthorization bills often contain STEM education-related provisions and at least 13 federal agencies conduct STEM education programs or activities. The federal investment in STEM education programs is estimated at between $2.8 billion and $3.4 billion annually.[6]

Congressional interest in STEM education heightened in 2007 when the National Academies published a report titled *Rising Above the Gathering Storm: Energizing and Employing America for a Brighter Economic Future*.[10] This influential publication warned federal policymakers that perceived weaknesses in the existing U.S. STEM education system—along with other important factors—threatened national prosperity and power. Although some analysts disputed its assertions, the report helped focus the federal conversation about STEM education and led, in part, to passage of the America Creating Opportunities to Meaningfully Promote Excellence in Technology, Education, and Science Act (or America COMPETES Act). Among other things, that act authorized STEM education programs at the National Science Foundation (NSF), National Oceanic and Atmospheric Administration (NOAA), Department of Energy (DOE), and Department of Education (ED).

What Is STEM?

Whether it is visas for foreign workers, scholarships for STEM majors, or funding for scientific research, the question of what we mean by the term "STEM" is central to the federal policy conversation. Some federal agencies, such as the NSF, use a broader definition of STEM that includes psychology and the social sciences (e.g., political science, economics) as well as the so-called core sciences and engineering (e.g., physics, chemistry, mathematics).[7] Others, including the Department of Homeland Security (DHS), U.S. Immigration and Customs Enforcement (ICE), use a narrower definition that generally excludes social sciences and focuses on mathematics, chemistry, physics, computer and information sciences, and engineering.[8] Some analysts argue that field-specific definitions such as these are too static and that definitions of STEM should focus on "an assemblage of practices and processes that transcend disciplinary lines and from which knowledge and learning of a particular kind emerges."[9]

Congress reauthorized the America COMPETES Act in 2010 (P.L. 111-358), thereby advancing it to the implementation phase of the policy cycle. Congress may opt to reauthorize the act in 2013, when many of its provisions will expire. In the meantime the federal conversation about STEM education continues in the budget and appropriations processes and in the various STEM education-related bills introduced each year.

Given policymakers' ongoing interest in establishing the scope and scale of federal STEM education effort, the first section of this report examines federal agencies, programs, and funding for STEM education. The second section examines the performance of the U.S. STEM education system and includes data and sources frequently cited in federal STEM education policy debates. The third section analyzes various STEM education policy issues and options, including those that relate to STEM education as a whole and those that are specific to the kindergarten-throughgrade-12 (K-12) and higher education systems. **Appendix A** and **Appendix B** contain links to sources of STEM education data and publications and to selected major legislation in federal STEM education policy history.

THE FEDERAL EFFORT IN STEM EDUCATION

At the request of Congress, four inventories of federal STEM education programs and activities have been published in recent years; two by the Government Accountability Office (GAO), one by the Academic Competitiveness Council (ACC),[11] and one by the National Science and Technology Council (NSTC).[12] The first GAO study, in 2005, found 207 distinct federal STEM education programs funded at about $2.8 billion in FY2004 (hereinafter this report is referred to as "GAO-2005").[13] In 2007, the ACC found 105 STEM education programs funded at about $3.1 billion in FY2006 (hereinafter this report is referred to as "ACC-2007").[14] A 2011 report by the NSTC identified 252 "distinct investments" in STEM education funded at about $3.4 billion in FY2010 (hereinafter this report is referred to as "NSTC-2011").[15] A second GAO study, published in 2012, reported 209 programs funded at about $3.1 billion in FY2010 (hereinafter this report is referred to as "GAO-2012").[16]

The discrepancies between these inventories indicate that establishing the federal effort in STEM education is complex and subject to methodological challenges. Differences between the inventories are due, in part, to the lack of a common definition of what constitutes a STEM education program or activity. Auditors have also found STEM education activities performed by science mission agencies difficult to capture because such activities tend to be fiscally and organizationally integrated into what are otherwise primarily scientific programs. Funding calculations and program identification become even more intricate when broad-purpose education programs with a STEM goal are considered (e.g., teacher training programs that focus on mathematics in addition to other fields such as reading). Finally, some estimates of federal STEM education activities depend on agency self-reporting, which is a less-reliable auditing method.

Despite these limitations, these four inventories reveal several general patterns in federal STEM education investments. The next two sections will discuss the inventories' findings by federal agency, population served, and program objective.

Federal Programs by Agency

Each of the four congressionally mandated inventories of the federal STEM education effort found that virtually all federal agencies administer STEM education programs. However, three agencies account for about four-fifths of federal funding for STEM education:

the National Science Foundation (NSF) and the Departments of Education (ED) and Health and Human Services (HHS).

As **Figure 1** shows, all four inventories found that about one-third of the federal investment in STEM education is appropriated to the NSF.

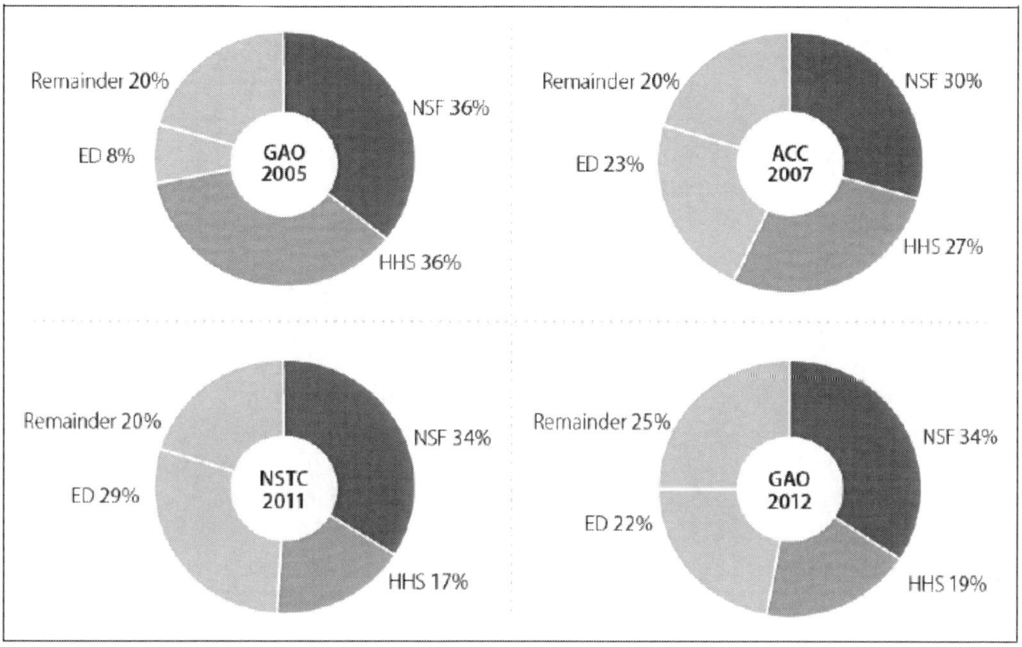

Source: CRS calculation based on GAO-2005, Figure 1; ACC-2007, Page 21; NSTC-2011, Figure 11; and GAO-2012, Appendix 2.

Figure 1. Federal STEM Education Funding, by Agency.

However, each inventory found different portions of STEM education funding at the other two agencies. The GAO-2005 and ACC-2007 inventories found a larger share of STEM funding at HHS than the GAO-2012 and NSTC-2011 studies. The GAO-2005 inventory found a much smaller share of funds at ED (8%; compared to 23%, 29%, and 22% in the latter three inventories). This discrepancy is likely attributable to a large increase in the FY2006 ED appropriation (roughly $310 million) for the National Science and Mathematics Access to Retain Talent (SMART) Grant program, which was newly authorized in 2005. Authority for the SMART Grant program ended in FY2010. No funds have been appropriated for the program since then.[17]

Population Served and Program Objectives

Each inventory took a different methodological approach and reported results somewhat incompatibly in terms of population served and program objective. This incompatibility is likely due to overlap between the populations served or program objectives within the individual STEM education programs. That is to say, sometimes the same program serves multiple populations (e.g., high school students and postsecondary students, graduate students and postdoctoral fellows). Additionally, nearly all STEM education programs have multiple objectives (e.g., supporting research and increasing degree attainment, encouraging advanced study and smoothing career transitions). Further, the inventories reported their findings on populations served and program objectives in different ways, thus making it difficult to compare their results on these important program elements.

Each inventory reported on programs by population served (e.g., by education level), although both GAO studies did this only in terms of the number of programs and not their funding level. Figure 2 presents the percentage of programs primarily serving elementary and secondary schools and postsecondary institutions as a share of the total number of programs identified in each inventory. Of the programs identified by GAO-2005, just fewer than 40% served elementary and secondary schools and just over 60% served postsecondary institutions; compared to 25% and 75% in the ACC-2007 study, 44% and 56% in the NSTC-2011 inventory, and 31% and 69% in the GAO-2012 study. The NSTC-2011 inventory did not include programs serving postdoctoral fellows, thus lowering the share of programs found at the postsecondary level.

Each inventory also collected information on program objectives. However, only the NSTC-2011 and GAO-2005 inventories reported information that could be summarized. The NSTC-2011 study found that nearly three-quarters of programs (74%) have at least two secondary program objectives in addition to the primary objective.[18] The GAO-2005 study found an even larger share of programs (80%) with more than one goal, with about half supporting four or more goals.[19]

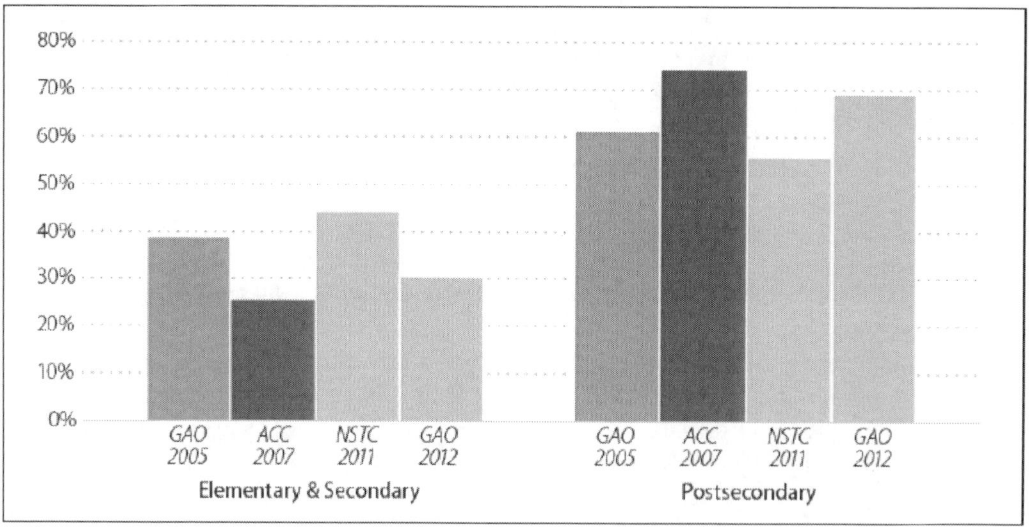

Source: CRS calculation based on GAO-2005, Table 8; ACC-2007, Page 2; NSTC-2011, Table 6; GAO-2012, Page 15.

Figure 2. Percentage of STEM Education Programs, by Education Level.

Figure 3 presents the share of federal STEM education programs by primary program objective for the GAO-2005 and NSTC-2011 inventories. In both studies, the majority of programs support degree attainment, research experience, and career development for postsecondary students (57% in the GAO study and 59% in the NSTC study). Fewer than one in five programs support STEM learning and engagement (GAO, 18%; NSTC, 13%). About one in ten programs support the training of STEM educators (GAO, 11%; NSTC, 9%).

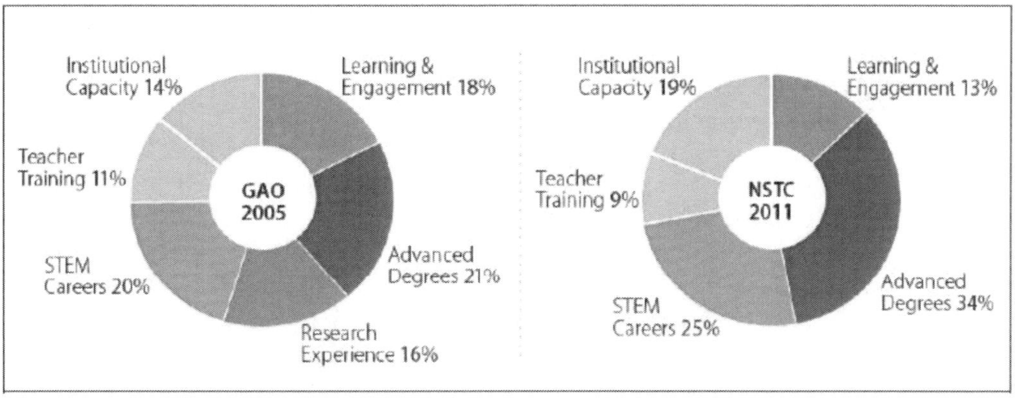

Source: CRS calculation based on GAO 2005, Table 6; NSTC-2011, Figure 7.

Figure 3. Percentage of STEM Education Programs, by Primary Objective.

Selected STEM Education Programs

In FY2012, the largest federal programs supporting STEM education were the Ruth L. Kirschstein National Research Service Awards ($274 million)[20] administered by HHS, the Graduate Research Fellowships program ($198 million) administered by NSF, and the Mathematics and Science Partnership program ($150 million) administered by ED. Not only are these the largest programs, they also represent two of the major activities receiving federal support, namely fellowships for graduate study and K-12 teacher training.

Ruth L. Kirschstein National Research Service Awards (HHS)

First funded in 1975, the Ruth L. Kirschstein National Research Service Awards (NRSA) constitute just under half (roughly 48%) of HHS spending on STEM education.[21] Most NRSA funds support Institutional Research Training Grants. About 15%-20% of funds support individual fellowships. The Institutional Research Training Grants are awarded to institutions to develop or enhance research training opportunities for individuals, selected by the institution, who are training for careers in specified areas of interest to the institution or principal investigator.[22] The individual fellowships are awarded directly to individuals from various organizations within the National Institutes of Health (e.g., the National Institute on Aging) to support the particular research interests of the individual receiving the award.[23]

Graduate Research Fellowships (GRF)

The Graduate Research Fellowship (GRF) program is the oldest and largest STEM education program at NSF. Established in 1952, the GRF is one of the most prestigious national awards offered to STEM graduate students. Fellows receive three-year portable stipends for graduate study leading to research-based master's and doctoral degrees in fields related to NSF's mission. Applicants are chosen by merit review. The NSF issued 7,800 fellowships (including 2,000 new fellowships) worth up to $42,000 each in FY2012. This amount includes a $12,000 cost-ofeducation (COE) allowance for the enrolling institution and a $30,000 stipend for the fellow.[24]

Some of the policy concerns associated with the GRF program focus on the number of fellowships offered annually, stipend and COE levels, and source(s) of funding.[25] Historically, funding for the GRF came primarily from NSF's main education account. Section 2 of the America COMPETES Reauthorization Act of 2010 (P.L. 111-358) directed NSF to fund the GRF equally from both the main education *and* research accounts. Funding for the GRF program increased after this change.[26] Some analysts propose expanding the GRF by creating a new NSF-industry fellows program, the funding for which would come equally from the private and public sectors.[27]

Mathematics and Science Partnerships (MSP)

The Mathematics and Science Partnership (MSP) program accounted for more than half (52%) of ED's STEM education portfolio in FY2012 ($150 million of $284 million). First authorized by the No Child Left Behind Act of 2001 (P.L. 107-110), the MSP program provides formula grants to states to increase the academic achievement of students in mathematics and science by enhancing the content knowledge and teaching skills of classroom teachers. With these funds, each State administers a grant competition in which awards are made to partnerships between high-need school districts and institutions of higher education. Grantees typically provide summer institutes and ongoing professional development designed to improve teachers' content knowledge through direct collaboration with scientists, mathematicians, and engineers.

In addition to ED's MSP, the 107th Congress created a companion program through the National Science Foundation Authorization Act of 2002 (P.L. 107-368). NSF's companion program is also called the Mathematics and Science Partnership (NSF-MSP) program.[28] Funded at $55 million in FY2012, NSF-MSP is a research and development effort that supports projects to serve as models of innovation for K-12 STEM education through competitive grants to institutions of higher education or nonprofit organizations in partnership with local education agencies. The NSF Director is required to report annually to Congress on how the program has been coordinated with ED's MSP program.[29]

THE CONDITION OF U.S. STEM EDUCATION

No single fact or statistic can wholly capture the condition of STEM education in the United States and for a variety of reasons the question "what is the condition of STEM education in the United States?" may be unanswerable. However, some trends appear to have held over time and in the most general sense, the condition of STEM education in the United States may be characterized as having more or less held constant or improved over the course of the last four decades. This is not the end of the story though. Looking at STEM education from this broad perspective disguises trends that concern many analysts and drive policy in this area. Among these concerns are persistent achievement gaps between various demographic groups, U.S. student performance on international mathematics and science tests, foreign student enrollments in U.S. institutions of higher education, global STEM education attainment, U.S. STEM teacher quality, and the U.S. STEM labor supply.

Upward Trends

According to the U.S. Census Bureau, the percentage of U.S. bachelor's degree holders with undergraduate degrees in science and engineering (S&E) was 36.4% in 2009 (approximately 20 million people).[30] This percentage is roughly consistent with the annual domestic production of S&E bachelor's degrees. The NSF estimates that the percentage of bachelor's degrees in S&E fields has held relatively constant—at between approximately 30% and 35% of all bachelor's degrees—for the past four decades. However, because the U.S. college-age population grew during these years, the total number of S&E bachelor's degrees awarded annually more than doubled between 1966 and 2008 (from 184,313 to 494,627).[31]

At the graduate level, S&E degrees predominate doctorate degree production. Since 1966, the percentage of doctorates in S&E fields has ranged between approximately 56% and 67% of all graduate degrees (where a field of study has been reported).[32] The total number of doctoral degrees in S&E fields has nearly tripled, growing from 11,570 in 1966 to 32,827 in 2008.[33] Graduate enrollments show similar upward trends. In 2010 there were 556,532 graduates enrolled in S&E fields (an historic peak), up from 413,536 in 2000.[34] Figure 4 displays the number of S&E degrees awarded, by degree level, over the last four decades.

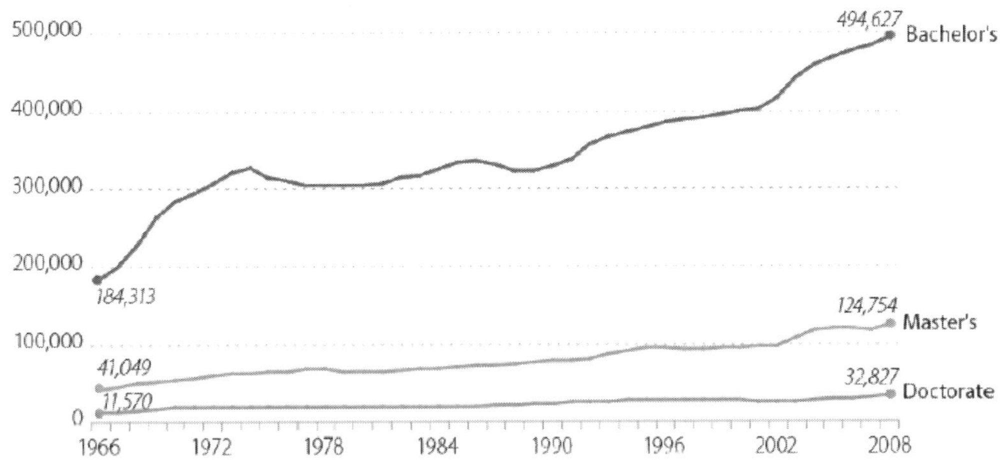

Source: National Science Foundation, National Center for Science and Engineering Statistics, "Table 1. S&E Degrees 1966-2008," *Detailed Statistical Tables* (NSF 11-316).

Notes: Includes only degrees where field of study is known. Includes degrees awarded in the social sciences and psychology.

Figure 4. Number of S&E Degrees Awarded from 1966-2008, By Degree Level.

Similar consistency in performance over time may also be found in the lower grades. The performance of U.S. K-12 students on standardized national mathematics tests has held constant or improved over the past four decades. For example, the average National Assessment of Educational Progress (NAEP) mathematics scores of students in 4th and 8th grades, grades in which students have been tested for decades, increased by 28 and 21 points—respectively— between 1990 and 2011.[35] **Figure 5** presents average NAEP math scores by various student subpopulations. Although all group scores have improved over time, sizable gaps remain.

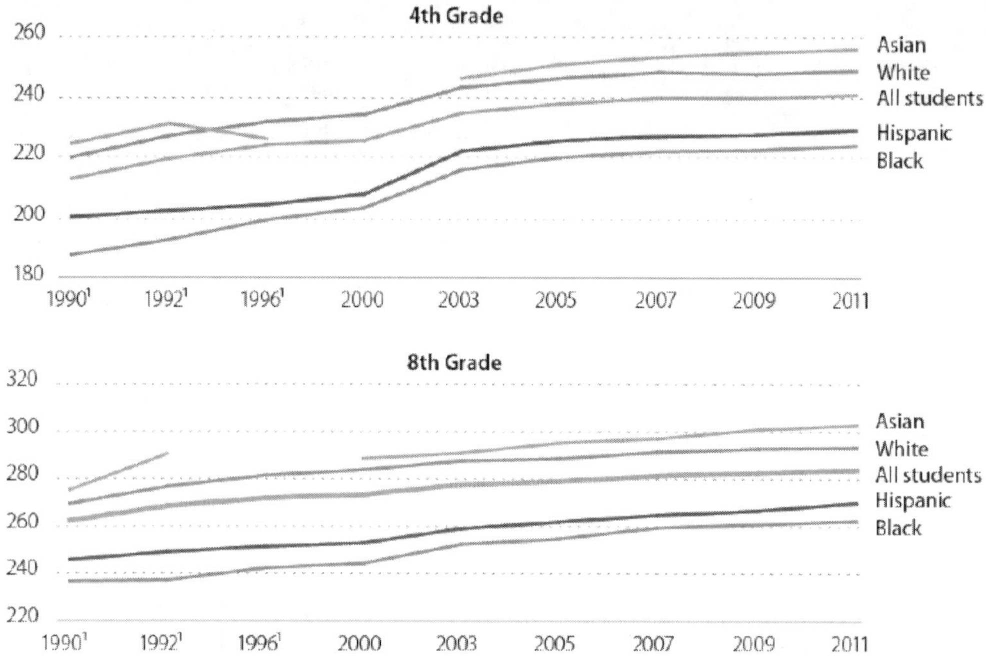

Source: CRS analysis of data from U.S. Department of Education, Institute of Education Sciences, National Center for Education Statistics, National Assessment of Educational Progress, various years.

Notes: The NAEP Mathematics scale ranges from 0 to 500. Some apparent differences between estimates may not be statistically significant. Time series are broken for years in which sample size was insufficient.

[1]Accommodations for students with disabilities were not permitted prior to 1996.

Figure 5. Trends in 4th and 8th Grade Average Mathematics Scores; Main NAEP, 1990 to 2011.

The average scores of 12th grade students on the main NAEP mathematics assessment were three points higher in 2009 than they were in 2005, when the test was first administered to this age group.[36]

Areas of Concern

In some respects, the overall trends paint a fairly optimistic picture for STEM education in the United States. Why, then, are so many observers so concerned about it? Analysts with concerns about STEM education cite a variety of data and trends as alarming. Among these are persistent achievement gaps between various demographic groups, U.S. student performance on international mathematics and science tests, foreign student enrollments in U.S. institutions of higher education, global STEM education attainment, U.S. STEM teacher quality, and the U.S. STEM labor supply.

Academic Achievement Gaps

A central topic in the conversation about STEM education focuses on so-called achievement gaps among various racial and ethnic groups and between women and men in certain STEM education outcomes. These gaps can be seen in a wide variety of STEM data, which show disparities by race, ethnicity, and gender in test scores, degree attainment, and employment. For example, there was at least a 20-point gap between the average scores of white students and their black and Hispanic counterparts on the 2011 4th and 8th grade NAEP mathematics assessments.[37] At the higher education level, only 18.5% of bachelor's degrees in engineering went to women in 2008.[38] Some STEM achievement gaps appear to hold relatively constant over time.

Although achievement gaps appear to be both pervasive and persistent, some evidence points to various types of improvement over time and in certain fields. For example, in the decade between 2000 and 2010, graduate enrollments in S&E fields grew by 35%. Further, among U.S. citizens and permanent residents, S&E graduate enrollments among Hispanic/Latino, American Indian/Alaska Native, and black/African America students grew at a higher rate than that of whites (not of Hispanic origin) and Asians.[39] While women account for relatively small percentages of degree recipients in certain STEM fields (as noted above), they accounted for 77.1% of the psychology degrees and 58.3% of the biological and agricultural sciences degrees in 2008.[40] Finally, although the 20+ point gap between the average scores of white students and their black and Hispanic counterparts on both the 4th and 8th grade NAEP mathematics tests has persisted for two decades, students of all ethnicities and races have higher average scores in 2011 than they did in 1990.[41]

Teacher Quality

Many observers look to the nation's teaching force as a source of shortcomings in student mathematics and science achievement. Research on teacher quality conducted over the last 20 years reveals that, among those who teach mathematics and science, having a major in the subject positively affects student achievement.[42] Unfortunately, many U.S. mathematics and science teachers lack this credential. For example, nearly all high school teachers have at least a baccalaureate degree; however, mathematics teachers are less likely than teachers of other subject areas to have majored in the subject they teach. In the 2007-2008 school year, roughly 17% of all high school teachers did not major in the subject they taught, while 28% of mathematics teachers did not major in mathematics.[43] Moreover, among those who majored in the subject they taught, mathematics teachers are less likely to be subject-certified than other teachers.

International Assessment Rankings

Another area often of concern is how U.S. students compare with their peers in other nations in their knowledge of mathematics and science. While U.S. students usually outscore the all-country average on international mathematics and science tests, they typically score below the average of industrialized nations. For example, U.S. 15 year-olds ranked below the Organization for Economic Cooperation and Development (OECD) average in mathematics—and ranked at the OECD average in science—on the 2009 Program for International Student Assessment (PISA).[44] U.S. students fare better on the Trends in International Mathematics and Science Study (TIMSS); U.S. 8th graders ranked 9th in mathematics and 11th in science on the 2007 TIMSS assessment.[45]

Many observers caution against using student assessments to compare nations. A variety of factors may influence test results, including translation issues, differences in test administration, student effort,[46] and the selection and diversity of test takers. The latter issue is often raised by critics of international assessments when looking at U.S. student performance. Some observers say that low performance in the United States is closely related to poverty, though the same reasoning applies to other countries. One analysis of the 2009 PISA results found that the richest U.S. areas (especially areas with less than 10% poverty) perform better than most other nations.[47]

Foreign Student Enrollment[48]

Although the number of degrees awarded in STEM fields has increased over time, many analysts are concerned about the percentage of STEM degrees that go to foreign students. For example, foreign students earn roughly one-third of all U.S. S&E doctoral degrees and earn half (or more) of U.S. doctoral degrees in the specific fields of engineering, physics, computer sciences, and economics.[49] Further, the percentage of doctoral degrees going to foreign students has been more or less increasing since the mid-1970s.[50]

The presence of foreign students in U.S. graduate S&E programs has been and continues to be of concern to some analysts because foreign graduates may not be eligible for work in the United States or for certain jobs requiring security clearance. Other observers suggest that these trends may mean missed opportunities or depressed wages for U.S. citizens and permanent residents who may be displaced by foreign graduates. Other analysts say that federal policymakers should encourage foreign STEM students to study and stay in the United States, arguing that policies meant to attract the world's best and brightest are key to ensuring U.S. competitiveness.[51]

Global STEM Education Attainment

In addition to concerns about foreign students obtaining STEM graduate degrees at U.S. institutions, some observers assert that the United States is falling behind other nations in the production of total STEM degrees. Of the 5 million first university degrees (e.g., undergraduate degrees) awarded globally in S&E in 2008, students in China earned about 23%, European Union students earned about 19%, and U.S. students earned about 10%. Further, while the United States awarded the largest number of total S&E doctoral degrees in 2008 (about 33,000), in 2007 China overtook the United States to become the world leader in the number of doctoral degrees awarded in the specific fields of natural sciences and engineering.[52]

Some analysts challenge these degree production numbers, arguing that at least in some cases (e.g., engineering) the United States produces higher quality graduates and that country-level comparisons are misleading because the statistics are not based on common methodologies or definitions.[53] However, attention to degree attainment trends has been amplified by scale differences between the sizes of the United States' and Chinese and Indian populations (i.e., about 300 million in the United States compared to about a 1.34 billion in China and 1.22 billion in India).

U.S. STEM Labor Supply[54]

A primary rationale behind federal STEM education policies relies on their perceived impact on the U.S. S&E workforce—and through it, on U.S. economic competitiveness and national security. Many business, academic, and policy leaders assert that U.S. STEM education weaknesses have contributed (or will soon contribute) to national S&E workforce shortages and that this labor supply problem has diminished U.S. global economic competitiveness and threatened national security (or will do so in the future).[55] However, some analysts argue that perceived limitations in the U.S. S&E workforce are overstated and that U.S. competitiveness is not threatened by across-the-board S&E labor shortages and does not require a supply-side response.[56] A third view holds that perceptions of S&E workforce shortages are accurate if the increasing numbers of jobs that are technically non-STEM, but that require STEM competencies (e.g., analytical skills), are included in labor demand calculations.[57]

Data and Methodological Limitations

Data are a big part of the current STEM education policy debate. Those who advocate for or against various STEM education policy proposals cite a variety of data and statistics in support of their assertions. However, in some cases data showing the impact of policy changes may lag behind those changes by years or decades, making accurate evaluation and policy assessment difficult. In other cases, data may be interpreted or used in ways that do not reflect potentially important research or methodological limitations. For example, one 2010 editorial stated that "the World Economic Forum ranked [the United States] 48th out of 133 developed and developing nations in quality of mathematics and science instruction."[58] The editorial did not explain that the source of the 48th place ranking was an opinion survey of global business executives. Although opinion surveys are often relevant in policy debate, policymakers may interpret their results differently than they would other kinds of evidence. These and other data limitations may challenge federal policymaking in this area.

STEM EDUCATION POLICY ISSUES

Stakeholders with an interest in improving STEM education suggest a wide and disparate set of policy options for Congress. Some of these recommendations address governance concerns about the administration of federal programs—e.g., removing duplication and improving program coordination within and across agencies. Other policy options focus attention on elements of the elementary and secondary school system—e.g., improving the quality and quantity of mathematics and science teachers and strengthening school accountability measures. Additional recommendations look to make improvements at the post-secondary level—e.g., enhancing retention of undergraduate STEM majors and strengthening incentives to pursue advanced STEM education. Many options focus on improving the STEM education outcomes of underrepresented populations.

Governance Concerns

Governance concerns are central to the contemporary debate about the federal STEM education effort. The scope and scale of the federal STEM education portfolio has some analysts concerned that federal agencies may be duplicating effort. In response to these concerns, some policymakers have proposed consolidating or eliminating STEM education programs. Other stakeholders support the broad and diffused nature of the federal STEM education effort and are more concerned with an apparent lack of coordination. Proponents of this view have argued for the development of an overarching federal STEM education strategy.

Duplication and Consolidation

Program consolidation is a widely debated policy option for federal STEM education programs. Advocates for this approach perceive duplication in the federal effort and assert that merging programs would result in cost savings. Proposals to consolidate STEM education programs have been made by both members of Congress and the Administration.[59] Some policymakers see program consolidation as a means to increase program flexibility and responsiveness because (under certain models of consolidation) federal program managers would have greater authority to shift priorities.[60] However, other policymakers may object to this change because it typically transfers program control from the legislative to the executive branch, shifting the balance of power.

Consolidation (particularly in the form of block grants) has also been proposed as a strategy to transfer control to the states and as a means to reduce program costs. Shifting control to the states could increase their ability to respond to local conditions and needs, but might make it more difficult to drive a national STEM education agenda[61] or to leverage unique assets of federal science agencies.[62] On the question of cost, the GAO has found that program consolidation can be more expensive in the short term and may not result in long term savings (if workloads are not also reduced) because administrative costs in federal STEM education programs tend to be less than 10% of total program costs.[63] Consolidation opponents raise general concerns about the potential impact of merging programs, arguing that certain programs (such as STEM education programs) need specified funding streams to avoid being passed over in favor of competing education priorities. It is unclear if this assertion would hold true in practice.

The impact of program consolidation on the federal STEM education effort would depend on what programs are consolidated, how the consolidation is accomplished, how funding streams are affected, and the degree to which programs are duplicative.[64] Congress could, for example, seek either a full or a partial consolidation of STEM education programs at individual agencies or across the entire portfolio. If Congress both consolidates programs and reduces funding levels, it may achieve savings from both administrative and program costs. Savings and program impacts from these changes would vary, depending on which of these strategies policymakers pursue.

Duplication or Overlap?

Published assessments of duplication in the federal STEM education portfolio are somewhat contradictory. Preliminary findings from April 2011 GAO testimony appeared to suggest the potential for duplication in federal teacher quality programs, including some STEM teacher programs.[65] However, the NSTC-2011 inventory specifically examined the duplication question within the federal STEM education portfolio and found "little overlap and no duplication."[66] The GAO-2012 inventory concluded that 83% of federal STEM education programs overlapped "to some degree," but stated that this overlap would "not necessarily be duplicative."[67]

Coordination and Strategy

Some stakeholders maintain that duplication in the federal portfolio is limited. They tend to focus instead on a perceived lack of coordination among and within agencies. To address this concern, some analysts call for an overarching STEM education strategy. Until recently, the federal STEM education effort was primarily undertaken in a distributed fashion that responded to the specific needs of agencies and STEM constituencies. That is, in general, programs were not part of a defined government-wide system with clear roles played by individual federal agencies. Some view this distributed approach as particularly sensitive to the unique workforce needs or STEM education assets of federal science agencies; other observers suggest that an overarching strategy may improve the efficiency of federal STEM education investments.[68]

Both the Congress and the President have moved to develop a federal STEM education strategy. Section 101 of the America COMPETES Reauthorization Act of 2010 (P.L. 111-358) directed the NSTC to develop and implement a five-year federal STEM education strategy. Although the NSTC had not published this strategy by mid-November 2012, it issued a status report in February 2012.[69] That status report identifies two common federal STEM education agency goals—STEM workforce development and STEM literacy—as well as policy and administrative strategies designed to accomplish these goals. In particular, the status report identifies four priority policy areas for the federal effort: "effective K-12 teacher education, engagement, undergraduate STEM education, and serving groups traditionally underrepresented in STEM fields."[70] The status report notes that strong arguments can be made for other STEM education policy areas, but states that these four were chosen as the priority areas for enterprise-wide coordination (agencies may still maintain their own STEM education priorities as well) because they represent the convergence of "national needs, Presidential priorities, and federal assets."[71]

To further enhance coordination at the federal level, some advocates maintain that Congress consider creating an Office of STEM Education and designating an Assistant Secretary for STEM Education at ED. Advocates for this approach claim that it would raise the profile of STEM education and improve administration of the various programs and policies at ED.[72]

Elementary and Secondary Schooling

Policymakers often express interest in making improvements in the early part of the STEM education pipeline—e.g., from kindergarten to 12th grade (also referred to as the "K-12" pipeline). Some analysts assert that mathematics and science achievement will not easily be raised unless the quality of K-12 teaching is improved. Other observers suggest low or unequal student achievement may be best addressed by adjustments to the K-12 system's accountability structure and standards for performance.

Teacher Quality

To many observers, mathematics and science teachers' lower likelihood of possessing subject-specific professional credentials, compared to teachers of other subjects, identifies a deficit of mathematics and science teacher quality. Although most teaching positions may be staffed, the K12 system's stock of fully credentialed mathematics and science teachers is in short supply.[73] A variety of solutions to the shortage of STEM teachers have been proposed.[74] One set of proposals is directed at teachers currently in the classroom, while another set of solutions targets new or prospective teachers.

Some advocates feel that it is important to focus on performance, and that current teachers who are less than fully effective in the classroom are not provided the support and training they need to succeed and want to see federal funding for professional development (PD) specifically designed for STEM teachers maintained and expanded.[75] Other stakeholders propose establishing a STEM Master Teacher Corps that would reward experienced and effective mathematics and science teachers with increased career prestige and pay in return for mentoring and providing PD for less effective teachers.[76] Still others support increased use of online education, especially in rural schools that struggle to attract new teachers in any subject.[77] Additionally, some researchers support reforms that would ease the removal of ineffective teachers who do not sufficiently improve with PD and may not be cut out for the profession.[78]

Other stakeholders think improved recruitment and retention of high-quality new teachers is the primary solution to the mathematics and science teacher quality problem. Many who take this approach argue that federal teacher policies should assist state and local efforts to improve nontraditional routes to teaching—e.g., alternative certification policies and incentives for mid-career transition of STEM professionals.[79] Analysts have identified options for attracting new STEM teachers through traditional preparation programs by subsidizing their education costs—e.g., through direct grants, student loan repayment, or tax credits[80]—and encouraging colleges and universities to develop concurrent STEM and teaching degree programs.[81] Such recruitment strategies may also serve as retention tools when paired with requirements that new teachers fulfill service agreements. Some proponents prefer policies designed to attract and retain STEM teachers through financial incentives such as differential pay, housing subsidies, and signing bonuses.

Alternatively, some education analysts have criticized the federal policy focus on teacher quality, as measured by credentials, calling into question its link to student achievement and advocating instead for proposals to improve teacher effectiveness.[82] Specifically, those in this camp suggest reforming teacher evaluation systems to identify multiple levels of effectiveness; rewarding those at the top with performance pay and removing those at the lowest level of performance.[83] Related proposals would reform the structure of teacher preparation by rewarding teacher training programs, both traditional and non-traditional, on the basis of their graduates' classroom effectiveness rather than on certification exam results.[84]

Accountability and Standards

The accountability movement has been a powerful force in the federal education policy debate for some time, reaching particular prominence with the passage of the No Child Left Behind Act of 2001 (NCLB, P.L. 107-110).[85] NCLB reforms were based on the premise that, to improve outcomes from the K-12 system, student achievement must be accurately assessed and schools must be held accountable for measurable results. The law required that states establish achievement benchmarks, set annual goals (referred to as Adequate Yearly Progress, or AYP), and have all students reach "proficiency" in reading and mathematics by 2014.[86] NCLB also requires that students be assessed for their academic proficiency in science. However, these results are not tied to the accountability system. Some stakeholders are in favor of amending the law to mandate that schools ensure students also be proficient in science.[87]

Independent of federal involvement, states have begun development of a so-called common core of academic standards across the K-12 system.[88] This effort intends to create nationally consistent standards of knowledge and skills that students need in order to graduate from high school and succeed in entry-level, credit-bearing college courses or workforce training programs. So far, standards have been developed for reading and mathematics and adopted by 45 states and the District of Columbia.[89] Pointing to the perceived "mediocre" state of current state science standards, some analysts say the inclusion of science in the common core "is long overdue."[90]

Other K-12 Policy Issues

Some analysts argue that the current "STEM for all" approach is not working.[91] Those in this camp urge policymakers to focus limited federal resources on high-achieving students with an interest in STEM by, among other things, using federal education funding to create new specialty STEM high schools.[92] Other analysts seek to expand programs such as Advanced Placement and International Baccalaureate (AP/IB)—including grants to pay the AP/IB testing fees of low-income students—or seek to increase STEM education achievement among demographic groups with historically low participation rates in STEM fields.[93] (See section on "Broadening Participation.")

Post-Secondary Education

As a proportion of all federal STEM education funding, the majority of the federal investment in STEM education supports undergraduate, graduate, and post-graduate education and research. In each of the recent program inventories, post-secondary education accounted for more than half of the federal STEM education portfolio. However, some analysts argue that current U.S. STEM degree production rates are insufficient.[94] Proposals to improve post-secondary STEM education include those that seek to address remediation in the early college years or increase retention rates in STEM majors through graduation. Other proposals seek increased support for graduate study and post-doctoral research. Some analysts favor lowering barriers for foreign STEM students seeking entry into U.S. institutions of higher education.

Remediation and Retention

Researchers cite poor pre-college mathematics and science preparation and high rates of attrition among STEM majors as two major challenges for undergraduate STEM education in the United States.[95] In addition to the K-12 improvements discussed above, some observers propose additional federal investment in remedial education for students as they enter college. For example, some stakeholders advocate for increased funding for ED's Upward Bound Mathematics and Science program.[96] Others analysts see community colleges playing an important role in counteracting the perceived failings of secondary schools. For example, some stakeholders have called for partnerships between business and two-year colleges to prepare students for STEM careers.[97] Other analysts argue that proprietary, non-degree-granting institutions are well suited to provide STEM remediation and training.[98]

Some policymakers are concerned with low retention rates at undergraduate STEM programs. Although attrition in STEM fields may be due, in part, to poor K-12 preparation and to overall college attrition patterns, there are likely multiple reasons why students complete a non-STEM degree after showing initial interest in STEM.[99] Some analysts advocate for STEM education research programs that focus on improving undergraduate STEM education practices for all students, such as NSF's Widening Implementation and Demonstration of Evidence-based Reforms (WIDER) and Transforming Undergraduate Education in Science, Technology, Engineering, and Mathematics (TUES) programs.[100] Others support efforts to improve retention among groups traditionally underrepresented in STEM fields (including ethnic and racial minorities, the disabled, and females).[101]

Broadening Participation of Underrepresented Populations

The demographic profile of the U.S. student-age population is changing. The youth population is more racially and ethnically diverse than previous generations of Americans. In addition, more than half of U.S. college students are now female, and over half of all bachelor's, master's, and doctoral degrees awarded in the United States go to women.[102] Some observers say that these trends are problematic for the U.S. scientific and technological enterprise, which has historically relied mostly on a white male labor supply (particularly in fields such as mathematics and engineering). However, because the growth in the student-age population (and therefore future labor supply) is in segments that have typically been underrepresented in STEM,[103] these observers argue that underrepresented groups "embody a vastly underused resource and a lost opportunity for meeting our nation's technology needs."[104] The solution to this challenge, many stakeholders argue, is to increase (or broaden) the participation of women and ethnic and racial minorities in STEM education and employment.[105]

General agreement about the problems posed by racial, ethnic, and gender disparities in STEM education and employment has not translated into widespread agreement on either the causes of underrepresentation or policy solutions. Further, causes and solutions may be different for different population subsets.

Race and Ethnicity

Researchers have identified dozens of school and non-school variables that may contribute to racial and ethnic achievement gaps in STEM. For example, in 2011 researchers reviewed over 400 books, book chapters, journal articles, and policy reports on factors that influence minority student success in STEM (hereinafter referred to as the "2011 review").[106] The 2011 review found that the following factors positively influence the success of minority students in STEM:

- *K-12*—parental involvement and support, availability of bilingual education, culturally relevant pedagogy, early exposure to STEM fields, interest in STEM careers, self-efficacy in STEM subjects,[107] and STEM-related educational opportunities and support programs.[108]

The 2011 review also identified the following school-based factors as contributing to minority under-preparedness in elementary and secondary STEM education:

- *K-12*—a lack of resources (underfunding) and less qualified teachers at schools that serve minority students, limited access to Advanced Placement courses, disproportionate tracking of minority students into remedial education, teachers' low expectations, stereotype threat,[109] racial oppression and oppositional culture,[110] and premature departure from high school.[111]

At the post-secondary education level, the 2011 review identified the following factors as associated with varying levels of minority student STEM success in college:

- *Higher Education*—colorblind meritocracy, financial aid and employment, institutional type, campus culture and climate, institutional agents, self-concept and self-efficacy, and STEM opportunity and support programs.[112]

In addition to these school-based factors described in the 2011 review, other researchers have identified non-school factors that contribute to achievement gaps in both STEM and non-STEM fields. Some scholars argue that these non-school factors have been overlooked and that too much emphasis is placed on schools.[113] Non-school factors that have been identified as contributing to achievement gaps include concentrated poverty and single-parent households,[114] early childhood development,[115] and health.[116]

Policy solutions for broadening participation in STEM are also numerous. In 2010, Congress directed the National Academies to examine diversity in the STEM workforce and make recommendations for broadening participation.[117] Of the many recommendations in the resulting report, the Academies identified several policy options as top priorities. These include increased financial support for minority undergraduate STEM students, improved teacher preparation, more and better advanced courses and academic advising for minority K-12 students, improved transition to graduate school for minority undergraduates in STEM fields, and increased availability of research assistantships for minority graduates students in STEM.[118] Other researchers have proposed solutions such as charter schools and school choice,[119] faith-based schooling,[120] improved transfer pathways from community colleges and reducing undergraduate debt,[121] and participation in undergraduate research.[122] Some analysts propose increased investments in minority serving institutions (MSIs)—such as Historically Black Colleges and Universities (HBCU) or Tribal Colleges and Universities (TCUs)—as an option to broaden participation in STEM fields.[123]

Gender

Although the number of women earning colleges degrees has been increasing, they hold less than a quarter of STEM jobs nationally.[124] Scholars debate the causes of gender disparities in STEM. Some analysts assert that self-efficacy, institutional culture, discrimination, and bias limit female participation in science.[125] Other observers do not find evidence of widespread, contemporary discrimination against women in STEM fields; instead, they primarily attribute disparities to family formation and child rearing, gendered expectations, lifestyle choices, career preferences, and personal choice, among other complex factors.[126]

Differences in beliefs about the causes of gender disparities in STEM lead to different emphases in proposed solutions. Scholars who generally align with the discrimination hypothesis suggest a variety of policy options. Among these are policies that seek to increase girls' interest in STEM; create college environments (e.g., institutional culture) that attract and retain female students and faculty; and counteract bias by, among other things, creating clear and transparent criteria for success.[127] Scholars who generally align with the preferences hypothesis recommend so-called family friendly policies at academic institutions (e.g., part-time tenure track positions and childcare) and propose federal funding for research "on the differing lifecourses of women's and men's careers to determine whether the traditional timing of hiring, tenure and promotion may deny society and science the contributions of talented women."[128]

Other Factors

Some researchers argue that income is the most critical variable in achievements gaps and that gaps between children from high- and low-income families have grown substantially in recent decades. The income achievement gap, these researchers argue, is as determinative (if not more) than race.[129] Researchers have identified summer learning loss as one of the possible contributors to achievement gaps by income.[130] Studies show that students lose skills over the summer, especially in mathematics, and that the effects of these losses appear to accumulate over time. Further, losses appear to disproportionately affect low-income students.[131]

Other researchers have observed mathematics achievement gaps by urbanization level (e.g., between rural, urban, and suburban youth), finding that urban and rural youth have lower average mathematics achievement levels than their suburban peers and that this gap appears to widen between kindergarten and 8th grade. These findings, say the researchers, are not solely attributable to differences in socio-economic status.[132]

APPENDIX A. DATA SOURCES AND MAJOR PUBLICATIONS

Federal STEM education analysts rely on a number of sources and major publications for data about the federal STEM education effort and the condition of STEM education in the

United States and around the globe. This appendix includes links to sources and publications where readers can find the most up-to-date STEM education data and information.

National and International Assessments

- **National Assessment of Educational Progress (NAEP)**—The NAEP is the largest nationally representative and continuing assessment of U.S. K-12 students. There are two NAEP assessments: Main NAEP and Long-Term Trends (LTT). The Main NAEP administers assessments in 12 subject areas, including mathematics and science. The LTT assesses mathematics and reading. More information about these assessments is available at
  - *Main NAEP Science*—http://nces.ed.gov/nationsreportcard/science/.
  - *Main NAEP Mathematics*—http://nces.ed.gov/nationsreportcard/ mathematics/.
  - *LTT Mathematics*—http://nationsreportcard.gov/ltt_2008/.
- **Trends in International Mathematics and Science Study (TIMSS)**—TIMSS is an international test that assesses the mathematics and science achievements of U.S. 4th and 8th grade students in a manner that is comparable across countries. More information about TIMSS is available at http://nces.ed.gov/timss/.
- **Program for International Student Assessment (PISA)**—PISA assesses the reading, mathematics, and science literacy of 15-year-old students in dozens of industrialized and developing nations. More information about PISA is available at http://nces.ed.gov/surveys/pisa/.

Federal Programs and Inventories

- U.S. Government Accountability Office, Science, Technology, Engineering, and Mathematics Education: Strategic Planning Needed to Better Manage Overlapping Programs Across Multiple Agencies (GAO-12-108)—This 2012 GAO report on federal STEM education programs includes an inventory of federal STEM education programs and policy recommendations. The report is available at http://gao.gov/products/GAO-12-108.
- National Science and Technology Council, A Report from the Federal Inventory of STEM Education Fast-Track Action Committee (NSTC 2011) and A Report from the Federal Coordination in STEM Education Task Force (NSTC 2012)—These reports provide an inventory of federal STEM education investments from FY2008 to FY2010 actual (in the 2011 report) and an update with information from FY2011 enacted to FY2013 requested (in the 2012 report).
  - NSTC 2011—This report is available at http://www.whitehouse.gov/sites/ default/files/microsites/ostp/ costem__federal_stem_education_portfolio_report.pdf.
  - NSTC 2012—This report is available at http://www.whitehouse.gov/sites/ default/files/microsites/ostp/ nstc_federal_stem_education_coordination_report.pdf.

- U.S. Department of Education, Academic Competitiveness Council (ACC), Report of the Academic Competitiveness Council—The 2007 ACC report provides an inventory of federal STEM education programs with funding data from FY2005 actual to the FY2007 President's budget request. Includes policy recommendations and an assessment of STEM education program evaluations from across the federal enterprise. The 2007 ACC report is available at http://www2.ed.gov/about/inits /ed/competitiveness/acc-mathscience/index.html.
- U.S. Government Accountability Office, Higher Education: Federal Science, Technology, Engineering, and Mathematics Programs and Related Trends (GAO-06-114)—This 2005 GAO report includes an inventory of federal STEM education programs and assesses program goals and constituencies served. The 2005 GAO report is available at http://www.gao.gov/products/GAO-06-114.

Condition of STEM Education

- **National Science Board, *Science and Engineering Indicators 2012***—This publication provides, among other things, one of the most comprehensive collections of key STEM indicators. It is published every two years. More information about *Science and Engineering Indicators* is available at http://www.nsf.gov/statistics /seind12/front/fronts6.htm.
- **National Science Foundation, National Center for Science and Engineering Statistics (NCSES)**—NCSES compiles and analyzes a variety of STEM data. Much of this data may be found in *Science and Engineering Indicators*, but the NCSES website includes separate, detailed, and timely publications on various STEM education data. More information about NCSES is available at http://www.nsf.gov/statistics/.

APPENDIX B. SELECTED MAJOR LEGISLATION

Depending on how broadly the term is defined, federal interest in STEM education may be traced to the 1st Congress. Several institutions that would become central parts of the federal STEM education effort—such as Health and Human Services (1798, 1 Stat. 605),[133] the Smithsonian Institution (1846, 9 Stat. 103), the National Academy of Sciences (1863, 12 Stat. 806), and Department of Education (originally the Office of Education, 1867, 14 Stat. 434)—were in place before the United States celebrated its first centennial.

Federal STEM education policymaking intensified after World War II. The desire to maintain the scientific achievements of the war led to the creation of the National Science Foundation in 1950. By 1952, NSF was issuing GRF awards to promising STEM graduate students. The Soviet Union's launch of the first artificial satellite, *Sputnik*, triggered fears that the United States was falling behind in mathematics and science education and led to the National Defense Education Act of 1958, which some analysts cite as the first federal foray into STEM education policy in the modern era.

This appendix includes selected historical federal STEM education measures arranged by date.[134]

Land Ordinance of 1785[135] and Northwest Ordinance of 1787[136]

The Land Ordinance of 1785 was one of a series of three measures providing for the political and geographic incorporation of the Northwest Territories in the Union. These measures were passed by the Continental Congress after the Revolutionary War and prior to the adoption of the Constitution. Drafted primarily for the purpose of disposing of land in the territories, the Land Ordinance of 1785 directed surveyors to establish townships in the territories. These townships were to be subdivided into lots, one of which (lot number 16) was to be preserved for the maintenance of a public school. The Land Ordinance's more famous cousin, the Northwest Ordinance of 1787, established governments in the territories and provided for the civil liberties of the inhabitants. On the question of education the Northwest Ordinance said, "Religion, morality and knowledge, being necessary to good government and the happiness of mankind, schools and the means of education shall forever be encouraged."[137]

Marine Hospital Service Act of 1798 (1 Stat. 605)[138]

Congress established the Marine Hospital Service (MHS) in 1798 to provide medical care for merchant seaman. Many federal health agencies trace their origin to the establishment of the MHS; including the National Institute of Health (NIH), which began as the Hygienic Laboratory within the MHS in 1887. The Ransdell Act of 1930 (P.L. 71-251) re-designated the Hygienic Laboratory as the NIH and authorized fellowships at the institute. Although NIH education and training funding in the 1930s, 1940s, and 1950s primarily focused on post-doctoral researchers and clinical traineeships for physicians, in 1948 the National Institute of Cancer began awarding funds to institutions to improve undergraduate education.[139] In 1974 Congress established the National Research Service Award (NRSA) at NIH. The National Research Service Award Act (P.L. 93-348) consolidated and established under a single authority existing research and fellowship authorities. P.L. 107-206 re-named the NRSA the "Ruth L. Kirschstein National Research Service Award" in 2002.

Morrill Acts of 1862 (12 Stat. 503)[140] and 1890 (26 Stat. 417)

The Morrill Act of 1862 authorized the sale of federal lands and distribution of the proceeds to the states for the purpose of establishing colleges in the "mechanic arts" (e.g., engineering, manufacturing, inventions), agriculture, and military tactics. The original Morrill Act did not apply to the "states in rebellion," but in 1890 Congress passed a subsequent measure to provide for colleges in Southern states.[141] The 1890 Morrill Act also expanded the purposes of the colleges to include "agriculture, the mechanic arts, the English language, and the various branches of mathematical, physical, natural, and economic science." These provisions were repealed in 1981 and replaced with "food and agricultural sciences."[142]

Colleges funded by these acts include the Massachusetts Institute of Technology, Clemson University, and many U.S. Historically Black Colleges and Universities.[143]

National Science Foundation Authorization Act of 1950 (P.L. 81-507)[144]

The NSF was established in 1950 to—in part—"develop and encourage the pursuit of a national policy for basic research and education in the sciences."[145] Congress passed the act authorizing the Foundation after several years of debate and a veto in 1947.[146] NSF distributed its first fellowships to pre- and post-doctoral STEM students in 1952. As early as 1953 the Foundation began supporting teacher institutes as a means of improving STEM education in the lower grades.[147] Although both the Congress and the President have made changes to the NSF since its founding, STEM education has remained a core function of the agency.

National Defense Education Act of 1958 (P.L. 85-864)

Passed in 1958 in response to the Soviet Union's launch of *Sputnik*, the National Defense Education Act (NDEA), sought to address concerns about "existing imbalances in our educational programs which have led to an insufficient proportion of our population educated in science, mathematics, and modern foreign languages and trained in technology."[148] Among its many provisions, the NDEA authorized the first federal student loan program; provided funds to states for science, mathematics, and modern foreign language instruction; and authorized grants to states for programs to identify and encourage gifted students. Some NDEA scholars assert that this act paved the way for the Elementary and Secondary Education Act of 1965 by establishing a legislative precedent for federal education aid to states.[149]

Elementary and Secondary Education Act of 1965 (P.L. 89-10)[150]

The primary source of federal aid to K-12 education is the Elementary and Secondary Education Act (ESEA).[151] ESEA was initially enacted in 1965 and was most recently amended and reauthorized by the No Child Left Behind Act of 2001 (NCLB, P.L. 107-110). STEM education was not central to the ESEA as originally constructed in 1965, but STEM-specific provisions have been added in subsequent reauthorizations. For example, as amended by No Child Left Behind, the act authorizes the Mathematics and Science Partnerships (MSP) program at ED[152] and requires states to have mathematics assessments and standards.

Higher Education Act of 1965 (P.L. 89-329)

The Higher Education Act (HEA) authorizes a series of programs that provide federal aid and support to institutions of higher education as well as a broad array of federal student aid programs that assist students and their families with paying for or financing the costs of obtaining a postsecondary education. The HEA was most recently amended in 2008 by the Higher Education Opportunity Act (P.L. 110-315); however, the only major STEM-related postsecondary program administered by ED was enacted by the Higher Education Reconciliation Act of 2005 (HERA; P.L. 109-171). The HERA amendments included authorization and appropriations for the SMART Grant program, which provided $4,000 grants to students majoring in STEM fields. Congress provided that the program sunset at the end of the 2010-2011 academic year. Approximately $1.4 billion in grants were awarded between FY2006 and FY2010.

Department of Education Organization Act of 1979 (P.L. 96-88)

The Department of Education Organization Act established ED as an independent federal agency. Section 304 of the act transferred science education programs established at NSF to ED.[153] Excluded from this directive were programs that related to scientific career development, continuing education of scientific personnel, career-focused broadening participation programs, research and development in science learning, and programs to inform the general public about the nature of science and technology and related policy issues. The conference report on the final bill included two specific examples of NSF programs to be transferred to ED: Elementary and Pre-school Science Teacher Training and Minority Institutions Science Improvement.[154] The act provided only for the transfer of programs in existence at the time of enactment and included a provision affirming NSF's authority to initiate and conduct programs under its originating act (P.L. 81-507).

Education for Economic Security Act of 1984 (P.L. 98-377)

The Education for Economic Security Act of 1984 (EESA) authorized teacher institutes and mathematics and science education development programs (including partnerships) at the National Science Foundation; directed the Department of Education to provide grants to states and local educational agencies for STEM teacher training and development; and authorized presidential awards for teaching in mathematics and science, among other things. EESA was enacted following publication of several reports—most notably *A Nation at Risk*—that were highly critical of the U.S. education system and amid growing concerns about international competitiveness in the wake of the 1970s recession and apparent ascendancy of the Japanese and German economies.

America COMPETES Act of 2007 (P.L. 110-69) and America COMPETES Reauthorization Act of 2010 (P.L. 111-358)[155]

The America COMPETES Act (and its 2010 reauthorization) authorized a variety of STEM education programs at several federal science agencies and ED. Most of the specified STEM education appropriations authorizations in these acts are at the NSF, but the acts also contain STEM education provisions for the Department of Energy (DOE), the National Aeronautics and Space Administration (NASA), and the National Oceanic and Atmospheric Administration (NOAA). Among other things, the 2010 reauthorization established a federal government-wide STEM education coordinating committee under the National Science and Technology Council.

End Notes

[1] Earlier examples include debate at the Constitutional Convention about whether to empower the federal government "to establish seminaries for the promotion of literature and the arts and sciences." James Madison, "Saturday, August 18," *Notes of Debates in the Federal Convention of 1787*, TeachingAmericanHistory.org website.

[2] U.S. President George Washington, *First Annual Message to Congress on the State of the Union, January 8, 1790*, The American Presidency Project website.

[3] The term "scientific thinking" has many definitions. In general, it refers to the skills, processes, and methods of science (broadly defined).

[4] Although a global competitiveness rationale drives much of the contemporary debate about STEM education policies, STEM illiteracy (particularly innumeracy) has also been linked to other national challenges such as the mortgage crisis and even medication errors. For example, see Kristopher Gerardi et al., *Financial Literary and Subprime Mortgage Delinquency: Evidence from a Survey Matched to Administrative Data*, Working Paper 2010-10, Federal Reserve Bank of Atlanta, April 2010; and Robert Preidt, "Parents' Poor Math Skills May = Medication Errors," National Institutes of Health, U.S. National Library of Medicine, Medline Plus website, April 30, 2012.

[5] Another source of STEM labor in the United States is immigration. For more information about foreign STEM workers, see CRS Report R42530, *Immigration of Foreign Nationals with Science, Technology, Engineering, and Mathematics (STEM) Degrees*, by Ruth Ellen Wasem.

[6] This is a rough estimate. The limitations of this calculation are explained in the section on "The Federal Effort in STEM Education."

[7] The America COMPETES Act of 2010 defines the term STEM for the agencies it authorizes, including the NSF. As defined by P.L. 111-358, Section 2, the term STEM means "the academic and professional disciplines of science, technology, engineering, and mathematics." In practice, NSF funds research in the so-called core sciences (e.g., mathematics and physical sciences) and engineering as well as psychology and the social sciences.

[8] While the DHS definition of a STEM field is, in general, more narrow than that of the NSF, DHS announced in May 2012 that it was expanding the list of fields it would support to include pharmaceutical sciences, econometrics, quantitative economics, and others. U.S. Department of Homeland Security, "DHS Announces Expanded List of STEM Degrees," press release, May 11, 2012. See also, U.S. Department of Homeland Security, Immigration and Customs and Enforcement, *STEM-Designated Degree Program List: 2012 Revised List*, 2012.

[9] Jean Moon and Susan Rundell Singer, "Bringing STEM into Focus," *Education Week*, vol. 31, no. 19 (February 1, 2012), pp. 32, 24.

[10] National Academy of Sciences, National Academy of Engineering, and Institute of Medicine, Committee on Prospering in the Global Economy of the 21st Century: An Agenda for American Science and Technology, and Committee on Science, Engineering, and Public Policy, *Rising Above the Gathering Storm: Energizing and Employing America for a Brighter Economic Future* (Washington, DC: National Academies Press, 2007).

[11] The ACC was created by the Deficit Reduction Act of 2005 (P.L. 109-171) and charged with conducting a yearlong study to identify all federal STEM education programs. U.S. Department of Education, *Report of the Academic Competitiveness Council*, May 2007.

[12] President Bill Clinton established the NSTC by Executive Order 12881 on November 23, 1993. The NSTC aims to coordinate science and technology policy across the federal government. For more information on the NSTC, see CRS Report RL34736, *The President's Office of Science and Technology Policy (OSTP): Issues for Congress*, by John F. Sargent Jr. and Dana A. Shea.

[13] U.S. Government Accountability Office, *Federal Science, Technology, Engineering, and Mathematics Programs and Related Trends*, GAO-06-114, Washington, DC, October 2005.

[14] In 2010, using a method similar to that of the ACC, the Office of Management and Budget (OMB) found 171 federal STEM education programs funded at about $3.8 billion. Unpublished data from the OMB. Available upon request.

[15] Executive Office of the President, National Science and Technology Council, Committee on STEM Education, Federal Inventory of STEM Education Fast-track Action Committee, *The Federal Science, Technology, Engineering, and Mathematics (STEM) Education Portfolio*, December 2011.

[16] U.S. Government Accountability Office, *Science, Technology, Engineering, and Mathematics Education: Strategic Planning Needed to Better Manage Overlapping Programs across Multiple Agencies*, GAO-12-108, January 2012.

[17] As enacted by the Higher Education Reconciliation Act of 2005 (P.L. 109-171), the SMART Grant program awarded $4,000 grants to students majoring in STEM fields. Congress provided that the program sunset at the end of the 2010- 2011 academic year. Approximately $1.4 billion in grants were awarded between FY2006 and FY2010.

[18] NSTC-2011, p. 16.

[19] GAO-2005, p. 13.

[20] This amount includes FY2012 enacted funding for Institutional Research Training Grants and individual fellowships. Executive Office of the President, National Science and Technology Council, Committee on STEM Education, Federal Coordination in STEM Education Task Force, *Coordinating Federal Science, Technology, Engineering, and Mathematics (STEM) Education Investments: Progress Report*, February 2012, p. 36.

[21] FY2012 funding for the NRSA is $273.5 million. Total STEM education funding at NIH, according to the NSTC-2011 inventory (updated), is $560.4 million. See Executive Office of the President, National Science and Technology Council, Committee on STEM education, Federal Coordination in STEM Education Task Force, *Coordinating Federal Science, Technology, Engineering, and Mathematics (STEM) Education Investments: Progress Report*, February 2012, p. 36.

[22] NRSA offers several types of Institutional Research Training Grants. Institutional eligibility varies.

[23] More information about the NRSA is available at http://grants.nih.gov/training/nrsa.htm.

[24] National Science Foundation, *FY2013 Budget Request to Congress*, February 13, 2012, pp. NSF-wide Investments-68-69. GRF program rules require institutions of higher education to exempt GRF fellows from tuition and fees. The COE provides funds to the institution for the cost of educating the student. The institution is responsible for tuition and fees in excess of the COE. Stipends are a form of salary and may be treated as taxable income.

[25] Ibid. NSF raised the COE from $10,500 to $12,000 in FY2012. For FY2013, the NSF seeks to increase the stipend level to $32,000. The FY2013 NSF budget request also seeks to increase the number of awards to 8,900 (2,000 new).

[26] Between FY2003 and FY2008, funds for the GRF came principally from NSF's main education account. Funding levels for the GRF during this period ranged between $85 million and $96 million. Starting in FY2009, the NSF increased the main research account contribution to the GRF program from less than $10 million per year to between $34 and $56 million per year. For FY2013 the NSF seeks a total of $243 million for the GRF. The main research and education accounts would each provide about half of this amount.

[27] Robert Atkinson et al., *Innovation Policy on a Budget: Driving Innovation in a Time of Fiscal Constraint*, Information Technology and Innovation Foundation, September 2010, p.10.

[28] In its report on legislation authorizing the MSP program at NSF, the House Committee on Science noted "The Committee believes that the Partnership program in this Act is complementary to, and not duplicative of a similarly titled math and science partnership program in H.R. 1, 'The No Child Left Behind Act of 2001.' ... The Committee anticipates that the two programs will draw on each others' strengths and that the most promising NSF-funded projects will be used as models and brought to full scale by the Department of Education's partnership program." See H.Rept. 107-134.

[29] For more information on NSF's MSP program, see National Science Foundation, Directorate for Education and Human Resources, *National Impact Report: Mathematics and Science Partnership Program*, NSF 2010-046, 2010, p. 18.

[30] Julie Siebens, *Science and Engineering Degrees: 2009*, U.S. Department of Commerce, Economics and Statistics Administration, U.S. Census Bureau, American Community Survey Briefs, ACSCR/09-14, October 2010.

[31] The low was 30.5% in 1991 and the high was 35.6% in 1968 and 1970. National Science Foundation, National Center for Science and Engineering Statistics, "Table 1," *S&E Degrees: 1966-2008*, NSF 11-316, June 2011.

[32] Ibid. The low was 56.1% in 1976 and the high was 67.5% in 2008.

[33] Ibid.

[34] Kelly Kang, NCSES *Infobrief: Graduate Enrollment in Science and Engineering Grew Substantially in the Past Decade but Slowed in 2010*, National Science Foundation, National Center for Science and Engineering Statistics, NSF 12-317, May 2012.

[35] Out of a possible 500 points, the average scores of 4th graders have gone from 213 in 1990 to 241 in 2011; 8th graders have gone from 263 to 284. The 1990 scores are statistically significant ($p < .05$) from 2011. U.S. Department of Education, Institute of Education Sciences, National Center for Education Statistics, *The Nation's Report Card: Mathematics 2011*, NCES 2012-458, November 2011; and, U.S. Department of Education, Institute of Education Sciences, National Center on Education Statistics, *2011 Condition of Education*, NCES 2011-033, May 2011, p. 46.

[36] Comparable data for the NAEP science assessment are not available. The science assessment was changed in 2009 to reflect changes in curriculum standards, assessments, research, and science. As such, the 2009 results are not comparable with results from prior years.

[37] The 2011 gap between the average scores of Hispanics and whites on the NAEP mathematics test was 20 points; for black and white students, the gap was 25 points. For more information on NAEP results and scoring, see U.S. Department of Education, Institute of Education Sciences, National Center for Education Statistics, *The Nation's Report Card: Mathematics 2011*, NCES 2012-458, November 2011, p. 13.

[38] National Science Foundation, National Center for Science and Engineering Statistics, "Table 11," *S&E Degrees: 1966-2008*, NSF 11-316, June 2011, p. 15.

[39] The rates for Hispanic/Latino, American Indian/Alaska Native, and black/African American S&E graduate enrollments between 2000 and 2010 were 65%, 55%, and 50%, respectively. Kelly Kang, NCSES *Infobrief: Graduate Enrollment in Science and Engineering Grew Substantially in the Past Decade but Slowed in 2010*, National Science Foundation, National Center for Science and Engineering Statistics, NSF 12-317, May 2012.

[40] National Science Foundation, National Center for Science and Engineering Statistics, "Table 11," *S&E Degrees: 1966-2008*, NSF 11-316, June 2011, p. 15.

[41] For more information on student achievement, see CRS Report R40514, *Assessment in Elementary and Secondary Education: A Primer*, by Rebecca R. Skinner.

[42] Michael B. Allen, *Eight Questions on Teacher Preparation: What Does the Research Say?*, Education Commission of the States, July 2003.

[43] U.S. Department of Education, Institute of Education Sciences, National Center for Education Statistics, "Table 2," *Education and Certification Qualifications of Departmentalized Public High School-Level Teachers of Core Subjects: Evidence from the 2007-08 Schools and Staffing Survey,* NCES 2011-317, May 2011, p. 14.

[44] Organization for Economic Cooperation and Development, *PISA 2009 Results: What Students Should Know and Can Do: Volume I*, December 2010.

[45] Ina V.S. Mullis et al., *TIMSS 2007 International Mathematics Report: Findings from IEA's Trends in International Mathematics and Science Study at the Fourth and Eighth Grades*, Boston College, Lynch School of Education, TIMSS & PIRLS International Study Center, 2008; and Michael O. Martin, et al., *TIMSS 2007 International Science Report: Findings from IEA's Trends in International Mathematics and Science Study at the Fourth and Eighth Grades*, Boston College, Lynch School of Education, TIMSS & PIRLS International Study Center, 2008.

[46] Some research has found that U.S. students do not try very hard on low-stakes standardized tests and that this affects scores. For example, research on financial incentives to improve student performance found that "The large effects of these relatively modest financial incentives [$10 to $20] suggest that at baseline this population of students [e.g., students in the study sample] puts forth low effort in response to low (perceived) returns to achievement on standardized tests." Steven D. Levitt et. al., *The Behavioralist Goes to School: Leveraging Behavioral Economics to Improve Educational Performance*, National Bureau of Economic Research, Working Paper no. 18165, June 2012.

[47] National Association of Secondary School Principals, "NASSP Responds to International Assessment Results," press release, December 10, 2010; and Cynthia McCabe, "The Economics Behind International Education Rankings," *NEAToday*, December 9, 2010.

[48] For more information on issues related to foreign students and foreign technical workers, see CRS Report R42530, *Immigration of Foreign Nationals with Science, Technology, Engineering, and Mathematics (STEM)*

Degrees, by Ruth Ellen Wasem; and CRS Report 97-746, *Foreign Science and Engineering Presence in U.S. Institutions and the Labor Force*, by Christine M. Matthews.

[49] In 2009, there were 611,629 graduate students in science and engineering fields in the United States. Of these, 168,850 (27.6%) were temporary residents. National Science Board, *Science and Engineering Indicators: 2012*, NSB 12-01, National Science Foundation, January 13, 2012, p. 2-28.

[50] National Science Foundation, "Figure 3.7—Citizenship Status of Ph.D.'s: 1960-1999," *U.S. Doctorates in the 20th Century*, NSF 06-319, October 2006; and Mark K. Feigener, *Number of Doctorates Awarded in the United States Declined in 2010*, National Science Foundation, NSF 12-303, November 2011.

[51] The House Committee on the Judiciary examined foreign student policy issues in an October 5, 2011, hearing titled, "STEM the Tide: Should America Try to Prevent and Exodus of Foreign Graduates of U.S. Universities with Advanced Degrees?" A video of the hearing, as well as written testimony from witnesses, is available at http://judiciary.house.gov/hearings/hear_10052011_2.html.

[52] These totals include foreign students. China expanded its domestic production of S&E doctoral degrees from about 2,700 in 1994 to almost 28,500 in 2008. National Science Board, *Science and Engineering Indicators, 2012*, NSB 12- 01, National Science Foundation, January 13, 2012, pp. 2-5 and 2-34 to 2-37.

[53] Gary Gereffi et al., "Getting the Numbers Right: International Engineering Education in the United States, China, and India," *Journal of Engineering Education*, vol. 97, no. 1 (January 2008), p. 13-25.

[54] For more information on issues related to the U.S. STEM labor supply, see CRS Report RL34091, *Globalization, Worker Insecurity, and Policy Approaches*, by Raymond J. Ahearn; CRS Report RL32292, *Offshoring (or Offshore Outsourcing) and Job Loss Among U.S. Workers*, by Linda Levine, CRS Report R42141, *Computer-Related Occupations Under the Fair Labor Standards Act (FLSA)*, by Gerald Mayer; and CRS Report R42411, *The Tool and Die Industry: Contribution to U.S. Manufacturing and Federal Policy Considerations*, by Bill Canis.

[55] Multiple reports from a variety of respected U.S. academic, scientific, and business organizations have made this argument. For example, see National Academy of Sciences, National Academy of Engineering, and Institute of Medicine, Committee on Prospering in the Global Economy of the 21st Century: An Agenda for American Science and Technology, and Committee on Science, Engineering, and Public Policy, *Rising Above the Gathering Storm: Energizing and Employing America for a Brighter Economic Future* (Washington, DC: National Academies Press, 2007); and, Frederick M. Hess, Andrew P. Kelly, and Olivia Meeks, *The Case for Being Bold: A New Agenda for Business in Improving STEM Education*, U.S. Chamber of Commerce, Institute for a Competitive Workforce, April 2011.

[56] For example, see Richard Freeman, "The Market for Scientists and Engineers," *NBER Reporter*, no. 3 (Summer 2007), pp. 6-8; Ron Hira, "U.S. Policy and the STEM Workforce System," *American Behavioral Scientist*, vol. 53, no. 7 (March 2010), pp. 949-961; Testimony of Institute for the Study of International Migration Director of Policy Studies B. Lindsay Lowell in U.S. Congress, House Committee on the Judiciary, Subcommittee on Immigration Policy and Enforcement, *"STEM" the Tide: Should America Try to Prevent an Exodus of Foreign Graduates of U.S. Universities with Advanced Science Degrees?*, hearings, 112th Cong., 1st sess., Serial No. 112-64, October 5, 2011; and B. Lindsay Lowell and Harold Salzman, *Into the Eye of the Storm: Assessing the Evidence on Science and Engineering Education, Quality, and Workforce Demand*, Urban Institute, October 2007.

[57] Anthony P. Carnevale, Nicole Smith, and Michelle Melton, *STEM: Science, Technology, Engineering, and Mathematics*, Georgetown University Center on Education and the Workforce, October 20, 2011, p.7.

[58] New York Times Editorial Board, "48th is Not a Good Place," *New York Times*, October 26, 2010, p. A28.

[59] For example, see Senator Tom Coburn, *The National Science Foundation: Under the Microscope*, April 2011, p. 54; and U.S. Department of Education, *A Blueprint for Reform: The Reauthorization of the Elementary and Secondary Education Act,* March 2010.

[60] This argument is, for example, part of the rationale for Administration-proposed program consolidations at ED. For more information, see CRS Report R41355, *Administration's Proposal to Reauthorize the Elementary and Secondary Education Act: Comparison to Current Law*, by Rebecca R. Skinner et al.

[61] This difficulty in driving a national agenda would depend on how the grants to states were structured. Federal policymakers could still drive a national STEM education agenda if they made receipt of consolidated program funds contingent on meeting certain defined national goals. However, some states may reject such efforts as overly prescriptive.

[62] For example, the National Aeronautics and Space Administration (NASA) has both unique workforce needs (e.g., astrobiologists) and unique assets that it can bring to the national STEM education effort (e.g., teaching from space).

[63] GAO states, "over 90% of STEM education programs that reported administrative costs estimated having administrative costs lower than 10% of their total program costs." U.S. Government Accountability Office, *Science, Technology, Engineering, and Mathematics Education: Strategic Planning Needed to Better Manage Overlapping Programs Across Multiple Agencies*, GAO-12-108, January 2012, p. 22.

[64] For example, programs that appear duplicative by some measures (e.g., target group) may have different intangible assets that could affect program implementation and outcomes. In this sense, they may not be *strictly* duplicative.

[65] U.S. Government Accountability Office, *Opportunities to Reduce Potential Duplication in Federal Teacher Quality Programs*, GAO-11-510T, April 13, 2011.

[66] Executive Office of the President, National Science and Technology Council, Committee on STEM Education, Fast-Track Action Committee on Federal Investments in STEM Education, *The Federal Science, Technology, Engineering, and Mathematics (STEM) Education Portfolio*, December 2011, p. 37.

[67] See U.S. Government Accountability Office, *Science, Technology, Engineering, and Mathematics Education: Strategic Planning Needed to Better Manage Overlapping Programs Across Multiple Agencies*, GAO-12-108, January 2012, p. 20.

[68] The NSTC-2011 inventory stated that "the primary issue [instead of duplication] is how to strategically focus the limited federal dollars available within the vast landscape of opportunity so they will have the most significant impacts possible in areas of national priority." (Executive Office of the President, National Science and Technology Council, Committee on STEM Education, Fast-Track Action Committee on Federal Investments in STEM Education, *The Federal Science, Technology, Engineering, and Mathematics (STEM) Education Portfolio*, December 2011, p. 37.) In its January 2012 report, GAO recommended not only that a federal STEM education strategy plan be drafted, but that NSTC should also develop policies to ensure compliance. In particular, the GAO recommended that the NSTC develop (1) guidance for agencies on how to incorporate STEM education efforts into agency performance plans; (2) a framework for how agencies will be monitored to ensure they collect and report on strategic plan goals; and (3) guidance to help agencies determine the types of evaluations that may be feasible and appropriate for different types of STEM education programs. Additionally, GAO recommended that the NSTC work with agencies to identify programs that might be candidates for consolidation or elimination. (U.S. Government Accountability Office, *Science, Technology, Engineering, and Mathematics Education: Strategic Planning Needed to Better Manage Overlapping Programs Across Multiple Agencies*, GAO-12-108, January 2012, p. 31.)

[69] Executive Office of the President, National Science and Technology Council, Committee on STEM Education, Federal Coordination in STEM Education Task Force, *Coordinating Federal Science, Technology, Engineering, and Mathematics (STEM) Education Investments: Progress Report*, February 2012.

[70] Ibid. p. 13.

[71] Ibid. p. 17.

[72] Triangle Coalition for Science and Technology Education, "Doing What's Best for Science, Technology, Engineering, and Mathematics Education," talking points, January 2011.

[73] Richard M. Ingersoll and David Perda, *The Mathematics and Science Teacher Shortage: Fact and Myth*, Consortium for Policy Research in Education, CPRE Research Report #RR-62, Philadelphia, PA, March 2009.

[74] For a discussion of teacher issues, see CRS Report R41267, *Elementary and Secondary School Teachers: Policy Context, Federal Programs, and ESEA Reauthorization Issues*, by Jeffrey J. Kuenzi.

[75] For example, the Triangle Coalition for Science and Technology Education advocates for maintaining current funding for ED's MSP program and increased funding for professional development support under ED's Teacher Quality State Grant program. Triangle Coalition for Science and Technology Education, "Doing What's Best for Science, Technology, Engineering, and Mathematics Education," talking points, January 2011.

[76] Letter from STEM Education Coalition to Senators Tom Harkin and Michael B. Enzi, June 20, 2011.

[77] Jerry Johnson and Marty Strange, *Why Rural Matters 2009: State and Regional Challenges and Opportunities*, The Rural School and Community Trust, October 30, 2009.

[78] Saba Bireda, *Devil in the Details: An Analysis of State Teacher Dismissal Laws*, Center for American Progress, June 3, 2010.

[79] National Council on Teacher Quality, *Tackling the STEM Crisis*, Summer 2009.

[80] For more information about higher education grants, loans, and tax benefits, see CRS Report R42446, *Federal Pell Grant Program of the Higher Education Act: How the Program Works, Recent Legislative Changes, and Current Issues*, by Shannon M. Mahan; CRS Report R40122, *Federal Student Loans Made Under the Federal Family Education Loan Program and the William D. Ford Federal Direct Loan Program: Terms and*

Conditions for Borrowers, by David P. Smole; and CRS Report R41967, *Higher Education Tax Benefits: Brief Overview and Budgetary Effects*, by Margot L. Crandall-Hollick.

[81] National Science Teachers Association, *NSTA Position Statement: Science Teacher Preparation*, July 2004.

[82] For example, see Dan Lips and Jena Baker McNeill, *A New Approach to Improving Science, Technology, Engineering, and Mathematics Education*, The Heritage Foundation, Backgrounder no. 2249, April 15, 2009.

[83] National Center on Performance Incentives, *Teacher Performance Pay: A Review*, November 2006.

[84] The Education Schools Project, *Educating School Teachers*, September 2006.

[85] More information on ESEA accountability can be found in CRS Report R41533, *Accountability Issues and Reauthorization of the Elementary and Secondary Education Act*, by Rebecca R. Skinner.

[86] More information assessment in ESEA can be found in CRS Report R40514, *Assessment in Elementary and Secondary Education: A Primer*, by Rebecca R. Skinner.

[87] NSTA Reports, "Should Science Count Toward AYP?," *National Science Teachers Association website—NSTA Web News Digest*, February 7, 2011.

[88] The Common Core State Standards Initiative is a state-led effort coordinated by the National Governors Association and the Council of Chief State School Officers. More information may be found at http://www.corestandards.org/.

[89] The National Research Council's Board on Science Education is currently developing a conceptual framework to guide the development of new science education standards. More information may be found at http://www7.nationalacademies.org/bose/Standards_Framework_FAQs.html.

[90] Lawrence S. Lerner et al., *The State of State Science Standards 2012*, Thomas B. Fordham Institute, January 31, 2012, p. 4-5.

[91] The "STEM for all" approach asserts that STEM competencies are central to contemporary work, life, and citizenship and that all U.S. students should have some mastery of these subjects and skills. In general, a "STEM for all" approach seeks to distribute STEM education resources widely, across all student skill levels, rather than targeting federal resources at high-achieving students.

[92] Robert D. Atkinson and Merrilea Mayo, *Refueling the U.S. Innovation Economy: Fresh Approaches to STEM Education*, Information Technology and Innovation Forum, December 7, 2010.

[93] Eric W. Robelen, "Latest Wave of STEM Schools Taps New Talent," *Education Week*, vol. 31, no. 3 (September 14, 2011), p.1, and published online under the title, "New STEM Schools Target Underrepresented Groups," *Edweek.org*.

[94] For example, the President's Council of Advisors on Science and Technology calculates that the United States will need, over the next decade, "approximately one million more college graduates in STEM fields than expected under current assumptions." Executive Office of the President, President's Council of Advisors on Science and Technology, *Engage to Excel: Producing One Million Additional College Graduates with Degrees in Science, Technology, Engineering, and Mathematics*, cover letter, February 2012.

[95] One-quarter of first-year college students were required to take remedial courses because they were not ready for college-level work. (Hart Research Associates, *One Year Out: Findings from a National Survey Among Members of the High School Graduating Class of 2010*, The College Board, August 18, 2011.) Two-fifths of students entering college intending to major in a STEM field complete a STEM degree. (Executive Office of the President, President's Council of Advisors on Science and Technology, *Engage to Excel: Producing One Million Additional College Graduates with Degrees in Science, Technology, Engineering, and Mathematics*, February 2012.)

[96] Council for Opportunity in Education, "2011 Capitol Hill Talking Points: Upward Bound Math-Science," talking points, 2011.

[97] Business Roundtable, *Taking Action for America: A CEO Plan for Jobs and Economic Growth*, March 1, 2012.

[98] Robin Wilson, "For-Profit Colleges Change Higher Education's Landscape," *The Chronicle of Higher Education*, February 7, 2010.

[99] For example, one widely cited study of STEM attrition found that poor teaching quality is a factor. (Elaine Seymour and Nancy M. Hewitt, *Talking About Leaving: Why Under-Graduates Leave the Sciences* (Boulder, CO: Westview Press, 1997). Other analysts cite the influence of grades on students' decisions to leave STEM majors. (Ben Ost, "The Role of Peers and Grades in Determining Major Persistence in the Sciences," *Economics of Education Review*, vol. 29, no. 6 (December 2010), pp. 923-934. Some observers assert that certain institutional practices, such as using introductory STEM courses to "weed out" or limit the number of students seeking STEM majors, contribute to perceived attrition challenges. (Jeffrey Mervis, "Weed-out Courses Hamper Diversity," *Science*, vol. 334, no. 6961 (December 2011), p. 1333.)

[100] Executive Office of the President, President's Council of Advisors on Science and Technology, *Engage to Excel: Producing One Million Additional College Graduates with Degrees in Science, Technology, Engineering, and Mathematics*, February 2012.

[101] For example, see National Academy of Sciences, National Academy of Engineering, and Institute of Medicine; Committee on Underrepresented Groups and the Expansion of the Science and Engineering Workforce Pipeline; Committee on Science, Engineering, and Public Policy; Policy and Global Affairs, *Expanding Underrepresented Minority Participation: America's Science and Technology Talent at the Crossroads* (Washington, DC: National Academies Press, 2011).

[102] Based on 2009 enrollment in four-year, two-year, and less-than-two-year Title VI eligible institutions. U.S. Department of Education, Institute of Education Sciences, National Center for Education Statistics, *Digest of Education Statistics: Table 196*, May 2011.

[103] Generally, analysts consider a demographic group to be "underrepresented" in STEM if the group's rate of participation in the STEM field is inconsistent with the group's presence in some broader population. For example, if women make up over half of all college students but are only a quarter of the engineering majors, then some observers would consider women to be underrepresented in college engineering enrollments.

[104] National Academy of Sciences, National Academy of Engineering, and Institute of Medicine; Committee on Underrepresented Groups and the Expansion of the Science and Engineering Workforce Pipeline; Committee on Science, Engineering, and Public Policy; Policy and Global Affairs, *Expanding Underrepresented Minority Participation: America's Science and Technology Talent at the Crossroads* (Washington, DC: National Academies Press, 2011), pp. 1-2. Also, although not specific to STEM, one 2009 report found that the U.S. Gross Domestic Product (GDP) could have been between two to four percent higher if the achievement gap between Latino/black and white students were narrowed. McKinsey & Company, Social Sector Office, *The Economic Impact of the Achievement Gap in America's Schools*, April 2009.

[105] For example, see David Beede et al., *Education Supports Racial and Ethnic Equality in STEM,* U.S. Department of Commerce, Economics and Statistics Administration, ESA Issue Brief #05-11, September 2011; Irwin Kirsch et al., *America's Perfect Storm: Three Forces Changing Our Nation's Future*, Educational Testing Service, Policy Information Center, Policy Evaluation and Research Center, January 2007; and National Academy of Sciences, National Academy of Engineering, and Institute of Medicine; Committee on Underrepresented Groups and the Expansion of the Science and Engineering Workforce Pipeline; Committee on Science, Engineering, and Public Policy; Policy and Global Affairs, *Expanding Underrepresented Minority Participation: America's Science and Technology Talent at the Crossroads* (Washington, DC: National Academies Press, 2011).

[106] Each of the terms and factors from the 2011 review (as summarized in this report) are described in greater detail in Samuel D. Museus et al., "Racial and Ethnic Minority Students' Success in STEM Education," *ASHE Higher Education Report,* vol. 36, no. 6 (January 2011).

[107] The term "self-efficacy" refers to a student's confidence in his or her ability to learn STEM subjects.

[108] Samuel D. Museus et al., "Racial and Ethnic Minority Students' Success in STEM Education," *ASHE Higher Education Report,* vol. 36, no. 6 (January 2011), p. viii.

[109] "Stereotype threat" is a theory developed by some social psychologists to explain the perceived effects of negative group stereotypes on the academic performance of the targets of those stereotypes. In essence, the theory asserts that when confronted with negative group stereotypes (e.g., girls are bad at math), individuals perform more poorly than they are otherwise capable of doing on intellectual tests. (Claude M. Steele and Joshua Aronson, "Stereotype Threat and the Intellectual Test Performance of African Americans," *Journal of Personality and Social Psychology*, vol. 69. no. 5 (1995), pp. 797-811.)

[110] "Oppositional culture" is a theory developed by some social scientists to explain the academic disengagement of black students. In essence, the theory postulates that black Americans have formed a culture that opposes mainstream values (as reaction to racial oppression and discrimination) and that this oppositional culture leads black Americans to devalue academic success because it is associated with "acting white." (Signithia Fordham and John Ogbu, "Black Students' School Success: Coping with the Burden of Acting White," *Urban Review*, vol. 18, no. 3 (December 1985), pp. 176-206.) Some researchers criticize oppositional culture theory, arguing that African Americans actually maintain more pro-school values than whites, but that they lack the material conditions that foster the development of skills, habits, and styles rewarded by teachers. (James W. Ainsworth-Darnell and Douglas B. Downey, "Assessing the Oppositional Culture Explanation for Racial/Ethnic Differences in School Performance," *American Sociological Review*, vol. 63, no. 4 (August 1998), pp. 536-553.) In reflecting on the debate about oppositional culture theory, the authors of the 2011 review conclude,

"this theory could plausibly be used to explain the negative educational outcomes of racial and ethnic minorities in K-12, particularly in STEM education." (Museus et. al., p. 37.)

111 Samuel D. Museus et al., "Racial and Ethnic Minority Students' Success in STEM Education," *ASHE Higher Education Report,* vol. 36, no. 6 (January 2011), pp. vii-viii.

112 Samuel D. Museus et al., "Racial and Ethnic Minority Students' Success in STEM Education," *ASHE Higher Education Report,* vol. 36, no. 6 (January 2011), p. viii.

113 James J. Heckman, *The American Family in Black and White: A Post-Racial Strategy for Improving Skills to Promote Equality*, National Bureau of Economic Research, Working Paper no. 16841, March 2011.

114 Irwin Kirsch et al., *America's Perfect Storm: Three Forces Changing Our Nation's Future*, Educational Testing Service, Policy Information Center, Policy Evaluation and Research Center, January 2007.

115 Educational Testing Service, Policy Evaluation and Research Center, *Positioning Young Black Boys for Educational Success*, Policy Notes: News from the ETS Policy Information Center, vol. 19, no. 3 (Fall 2011); and Eugene E. Garcia et al., "Early Academic Achievement of Hispanics in the United States: Implications for Teacher Preparation," *New Educator*, vol. 2, no. 2 (2006), pp. 123-147.

116 Kevin Fiscella and Harriet Kitzman, "Disparities in Academic Achievement and Health: The Intersection of Child Education and Health Policy," *Pediatrics*, vol. 123, no. 3 (March 2009).

117 P.L. 110-69, Section 7032.

118 National Academy of Sciences, National Academy of Engineering, and Institute of Medicine; Committee on Underrepresented Groups and the Expansion of the Science and Engineering Workforce Pipeline; Committee on Science, Engineering, and Public Policy; Policy and Global Affairs, *Expanding Underrepresented Minority Participation: America's Science and Technology Talent at the Crossroads* (Washington, DC: National Academies Press, 2011).

119 Matthew Ladner and Lindsey Burke, *Closing the Racial Achievement Gap: Learning from Florida's Reforms*, The Heritage Foundation, Backgrounder no. 2468, October 4, 2010.

120 The White House Domestic Policy Council and U.S. Department of Education, *Preserving a Critical National Asset: America's Disadvantaged Students and the Crisis in Faith-Based Urban Schools*, September 2008, p. 7.

121 Alicia C. Dowd, Lindsey E. Malcolm, and Elsa E. Macias, *Improving Transfer Access to STEM Bachelor's Degrees at Hispanic Serving Institutions through the America COMPETES Act*, University of Southern California, Rossier School of Education, Center for Urban Education, March 2010; and Steve Olsen and Jay B. Labov, rapporteurs, *Community Colleges in the Evolving STEM Education Landscape*, National Research Council and National Academy of Engineering; Division on Policy and Global Affairs, Board on Higher Education and Workforce; Division on Earth and Life Studies, Board on Life Sciences; Division on Behavioral and Social Sciences and Education, Board on Science Education; National Academy of Engineering, Engineering Education Program Office; and Division on Behavioral and Social Sciences and Education, Teacher Advisory Council; Planning Committee on Evolving Relationships and Dynamics Between Two- and Four-Year Colleges, and Universities (Washington, DC: National Academies Press, May 8, 2012); and Alicia C. Dowd and Lindsey E. Malcom, *Reducing Undergraduate Debt to Increase Latina and Latino Participation in STEM Professions*, University of Southern California, Rossier School of Education, Center for Urban Education, May 2012.

122 Gina A. Garcia and Sylvia Hurtado, *Predicting Latina/o STEM Persistence at HSIs and Non-HSIs*, University of California, Los Angeles, Graduate School of Education and Information Studies, Higher Education Research Institute, April 2011.

123 For example, a March 16, 2010, congressional hearing on broadening participation in STEM included testimony on the role of MSIs in producing minority STEM graduates. U.S. Congress, House Committee on Science and Technology, Subcommittee on Research and Science Education, *Broadening Participation in STEM*, hearings, 111th Cong., 2nd sess., Serial No. 111-85 (Washington, DC: GPO, 2010).

124 U.S. Department of Commerce, Economics and Statistics Administration, *Women in STEM: A Gender Gap to Innovation*, ESA Issue Brief #04-11, August 2011.

125 Catherine Hill, Christianne Corbett, and Andresse St. Rose, *Why So Few: Women in Science, Technology, Engineering, and Mathematics*, American Association of University Women, February 2010; and National Academy of Sciences, National Academy of Engineering, and Institute of Medicine, Committee on Maximizing the Potential of Women in Academic Science and Engineering, *Beyond Bias and Barriers: Fulfilling the Potential of Women in Academic Science and Engineering* (Washington, DC: National Academies Press, 2007).

126 Stephen J. Ceci and Wendy M. Williams, "Understanding Current Causes of Women's Underrepresentation in Science," *Proceedings of the National Academy of Sciences*, vol. 108, no. 8 (February 7, 2011), p. 3157-3162.

[127] Corinne A. Moss-Racusin, et al, "Science Faculty's Subtle Gender Biases Favor Male Students," *Proceedings of the National Academy of Sciences*, vol. 109, no. 41 (October 9, 2012), p. 16474-16479; and Catherine Hill, Christianne Corbett, and Andresse St. Rose, *Why So Few: Women in Science, Technology, Engineering, and Mathematics*, American Association of University Women, February 2010, pp. 90-96.

[128] Stephen J. Ceci and Wendy M. Williams, "Understanding Current Causes of Women's Underrepresentation in Science," *Proceedings of the National Academy of Sciences*, vol. 108, no. 8 (February 7, 2011), p. 3161.

[129] Sean Reardon, "The Widening Achievement Gap Between the Rich and the Poor: New Evidence and Possible Explanations," *Whither Opportunity*, ed. Greg J. Duncan and Richard J. Murnane (New York, NY: Russell Sage Foundation, September 2011), p. 91.

[130] Other factors associated with summer learning loss include demographic characteristics. For example, one 2006 study found that high-performing African-American and Hispanic students lost more achievement than their European-American peers over the summer and that low-performing African-American and Hispanic students grow less than low-performing students in all groups. See Martha S. McCall et. al., *Achievement Gaps: An Examination of Differences in Student Achievement and Growth*, Northwest Evaluation Association, November 2006.

[131] Jennifer Sloan McCombs et al., *Making Summer Count: How Summer Programs Can Boost Children's Learning*, RAND Corporation, 2011.

[132] Suzanne E. Graham and Lauren E. Provost, *Mathematics Achievement Gaps Between Suburban Students and Their Rural and Urban Peers Increase Over Time*, University of New Hampshire, Carsey Institute, Issue Brief No. 52, Summer 2012.

[133] The Department of Health and Human Services (HHS) traces its history to establishment of the federal Marine Hospital Service, forerunner of the contemporary U.S. Public Health Service, in 1798. More information about HHS history is available at http://www.hhs.gov/about/hhshist.html.

[134] Unless otherwise indicated, historical STEM education measures in this section are described *as originally passed*. Most of these measures have been amended, in some cases quite significantly (including repeal), since they became law.

[135] John C. Fitzpatrick, ed., *Journals of the Continental Congress 1774-1789*, edited from the original records in the Library of Congress, vol. 28 (Washington, DC: GPO, 1933), p. 375, http://memory.loc.gov/cgi-bin/query/r?ammem/ hlaw:@field%28DOCID+@lit%28jc0281%29%29.

[136] Library of Congress, *Primary Documents in America History: Northwest Ordinance*, Library of Congress/Virtual Services Digital Reference Section/Web Guides website, April 25, 2012, http://www.loc.gov/rr/program/bib/ourdocs/ northwest.html.

[137] Library of Congress, *An Ordinance for the Government of the Territory of the United States, North-West of the River Ohio*, Library of Congress/American Memory/Documents from the Continental Congress and Constitutional Convention 1774-1789 website, no date, http://memory.loc.gov/cgi-bin/query/r?ammem/ bdsdcc:@field(DOCID+@lit(bdsdcc22501)).

[138] The statute establishing the MHS does not include a formal title for the act. For the sake of consistency with other headings in this section, CRS used the title "Marine Hospital Service Act of 1789" to describe 1 Stat. 605. A copy of this statute is available at http://memory.loc.gov/cgi-bin/ampage?collId=llsl&fileName=001/llsl001.db&recNum=728.

[139] U.S Congress, Senate Committee on Labor and Public Welfare, *National Research Service Award Act*, report to accompany H.R. 7724, 93rd Cong., 1st sess., S. Rept. 93-381.

[140] National Archives and Records Administration, *Morrill Act (1862)*, National Archives and Records Administration/100 Milestone Documents website, no date, http://www.ourdocuments.gov/doc.php?doc=33, http://www.ourdocuments.gov/doc.php?doc=33.

[141] The 1890 measure required states that accepted funds to either (a) discount race in admissions, or (b) provide separate colleges for white and black students.

[142] Agriculture and Food Act of 1981 (P.L. 97-98).

[143] Library of Congress, *Primary Documents in America History: Morrill Act*, Library of Congress/Virtual Services Digital Reference Section/Web Guides website, July 30, 2010, http://www.loc.gov/rr/program /bib/ourdocs/ Morrill.html.

[144] For more information on STEM education at the NSF, see CRS Report R42470, *An Analysis of STEM Education Funding at the NSF: Trends and Policy Discussion*, by Heather B. Gonzalez.

[145] National Science Foundation Act of 1950 (P.L. 81-507), http://www.nsf.gov/about/history/legislation.pdf.

[146] Controversy over the founding of the NSF focused mostly on organizational questions, concerns about patents, and on other issues not related to STEM education.

[147] Dorothy Schaffter, *The National Science Foundation* (New York, NY: Frederick A. Praeger, Publishers, 1969), p. 96.

[148] National Defense Education Act (P.L. 85-864).

[149] For example, see Barbara Barksdale Clowse, *Brainpower for the Cold War: The Sputnik Crisis and National Defense Education Act of 1958* (Westport, CT: Greenwood Press, 1981), p. 147; and Wayne J. Urban, *More Than Science and Sputnik: The National Defense Education Act of 1958* (Tuscaloosa, AL: University of Alabama Press, 2010), p. 202.

[150] For more information on ESEA, see CRS Report RL33960, *The Elementary and Secondary Education Act, as Amended by the No Child Left Behind Act: A Primer*, by Rebecca R. Skinner.

[151] Particularly Title I, Part A, Program of Education for the Disadvantaged.

[152] NSF hosts a companion program that is also called Mathematics and Science Partnerships (MSP). As currently authorized, the two programs were designed to complement each other.

[153] Department of Education Organization Act (P.L. 96-88), http://www.eric.ed.gov/PDFS/ED180121.pdf.

[154] U.S. Congress, Senate Committee on Governmental Affairs, *Legislative History of P.L. 96-88, Department of Education Organization Act* (Part 2), committee print, 96th Cong., 2nd sess. (Washington, DC: GPO 1980), p. 1758.

[155] For more information about the America COMPETES Act see CRS Report R41819, *Reauthorization of the America COMPETES Act: Selected Policy Provisions, Funding, and Implementation Issues*, by Heather B. Gonzalez.

In: Science, Technology, Engineering and Math …
Editor: Noa Lemoine
ISBN: 978-1-62808-690-4

Chapter 2

FEDERAL SCIENCE, TECHNOLOGY, ENGINEERING, AND MATHEMATICS (STEM) EDUCATION: 5-YEAR STRATEGIC PLAN*

Committee on STEM Education

Executive Office of the President
National Science and Technology Council
Washington, D. C. 20502
May 31, 2013

Members of Congress:

I am pleased to transmit this 5-Year Federal Science, Technology, Engineering, and Mathematics (STEM) Education Strategic Plan in response to the requirements of the America Competes Reauthorization Act of 2010. Prepared by the Committee on STEM Education (CoSTEM) of the National Science and Technology Council, it follows up the interim report that CoSTEM provided to you in February 2012 under the title *Coordinating Federal STEM-Education Investments: Progress Report.*

The health and longevity of our Nation's, citizenry, economy and environmental resources depend in large part on the acceleration of scientific and technological innovations, such as those that improve health care, inspire new industries, protect the environment, and safeguard us from harm. Maintaining America's historical preeminence in the STEM fields will require a concerted and inclusive effort to ensure that the STEM workforce is equipped with the skills and training needed to excel in these fields.

President Obama believes that now is a time of opportunity – that we can build a true "all hands on deck" effort to move America forward and address this challenge. During the President's first term, the Administration used multiple strategies to make progress on improving STEM education:

---

* This report was released by the National Science and Technology Council, May 2013.

- *Making STEM a priority in more of the Administration's education efforts.* The first round of the Department of Education's $4.3 billion Race to the Top competition offered states a competitive preference priority on developing comprehensive strategies to improve achievement and provide rigorous curricula in STEM subjects; partner with local STEM institutions, businesses, and museums; and broaden participation of women and girls and other groups underrepresented in STEM fields. Other examples include STEM priorities in the Department of Education's Invest in Innovation and Supporting Effective Educator Development programs. Prioritizing STEM in existing programs at the Department of Education has the advantage of leveraging existing resources and embedding STEM within our overall education reform efforts.
- *Setting ambitious but achievable goals and challenging the private sector.* President Obama announced the goal to prepare 100,000 excellent STEM teachers over the next decade in his 2011 State of the Union Address. Answering this call to action, over 150 organizations led by the Carnegie Corporation of New York formed a coalition called *100Kin10*. Members of the coalition have made over 150 commitments to support STEM-teacher preparation and have raised over $30 million for this effort. In mid-March, the Howard Hughes Medical Institute announced a $22.5M investment to support expansion of the successful UTeach program in support of this goal.[1] Additional examples of this all-hands-on-deck approach to challenging companies, foundations, non-profits, universities, and skilled volunteers include *Change the Equation*, US2020, and the scaling up and expanding an AP program for children in military families.
- *Capitalizing on the President's deep interest and leadership.* The President hosted the first-ever White House Science Fair in late 2010 and the second in 2012, fulfilling a commitment he made at the launch of his ***Educate to Innovate*** campaign to directly use his bully pulpit to inspire more boys and girls to excel in mathematics and science. The President has also issued a call to action to the 200,000 Federal scientists and engineers to volunteer in their local communities and think of creative ways to engage students in STEM subjects.

Improving STEM education will continue to be a high priority in President Obama's second term. Guided by the aims articulated in the February 2012 *Progress Report* and subsequent pre-final drafts of this Strategic Plan—as well by the President's desire to re-organize STEM-education programs for greater coherence, efficiency, ease of evaluation, and focus on his highest priorities—the Executive Office of the President recommended, and the President accepted, a FY2014 Budget Request for STEM education that would increase the total investment in STEM-ed programs by 6 percent over the 2012 appropriated level while reducing the *number* of programs spread across the 14 CoSTEM agencies from 226 to 110. The 78 programs that would be eliminated—with 38 others consolidated into the remaining 110— accounted in FY2012 for $176 million of the spending by the nine agencies losing programs, amounting to 15 percent of their STEM-ed spending in that year. The savings would be redirected to flagship programs in the Department of Education and the National Science Foundation, plus a new one at the Smithsonian Institution, as follows:

- The Department of Education will play an increased role in improving P-12 STEM instruction by supporting partnerships among school districts and universities, science agencies, businesses, and other community partners to transform teaching and learning. It will invest an additional $80 million in support of the 100,000 new STEM-ed teachers goal and $35 million for the launch of a pilot STEM-ed Master Teacher Corps, as well as in creation of new STEM Innovation Networks to better connect school districts with local, regional, and national STEM resources. The Department will also collaborate with all of the CoSTEM agencies to ensure that Federal scientific assets are utilized in the improvement of P-12 STEM education.
- The National Science Foundation will increase its focus on improving the delivery of undergraduate STEM teaching and learning through evidence-based reforms, including a new $123 million program aimed at improving retention of undergraduates in STEM fields. NSF will also receive $325 million to expand and enhance its graduate fellowship programs, including creation of a new National Graduate Research Fellowship, using a common infrastructure at NSF to reach more students and offer a set of opportunities that address national needs and mission-critical workforce needs for the CoSTEM agencies.
- The Smithsonian Institution will receive $25 million to focus on improving the reach of informal STEM education by ensuring that materials are aligned to what students are learning in the classroom. The Smithsonian will work with NSF, ED, the other CoSTEM agencies including the National Aeronautics and Space Administration (NASA), National Oceanic and Atmospheric Administration (NOAA), U.S. Department of the Interior (DOI), U.S. Department of Agriculture (USDA), National Institutes of Health (NIH), and other science partners to harness their unique expertise and resources to disseminate relevant, evidence-based materials and curricula, on-line resources, and delivery and dissemination mechanisms to reach more teachers and students both inside and outside the classroom.

All of the CoSTEM agencies will continue to be key players in the re-organized effort. All of these agencies depend upon the cultivation of a talented and well-trained workforce in order to meet their STEM-related missions, and all of them play a critical role in inspiring and training the next generation of STEM workers. Whether it be through direct support, provision of expertise and content, mobilization of talented STEM role models and mentors, or by exposing students to real-world learning opportunities at Federal STEM facilities, these agencies inspire and inform future scientists, engineers, innovators, and explorers.

The Strategic Plan I am transmitting today complements the important steps we have already taken. The Plan begins by providing an overview of the importance of STEM education to American scientific discovery and innovation, the need to better prepare students for today's jobs and those of the future, and the importance of a STEM-literate society (section 2) and also describes the current state of Federal STEM education efforts (section 3). The document then presents five priority STEM education investment areas where a coordinated Federal strategy can be developed, over five years, designed to lead to major improvements in key areas. This increased coordination is expected to bring significant gains in efficiency and coverage (section 4).

Also included in this plan are initial implementation roadmaps in each of the priority STEM education investment areas, proposing potential short-, medium-, and long-term objectives and strategies that might help Federal agencies achieve the outlined goals (section 5). Additionally, throughout the document, the plan highlights (1) key outcomes for the Nation and ways Federal agencies can contribute, (2) areas where agencies will play lead roles, thereby increasing accountability, (3) methods to build and share evidence, and (4) approaches for decreasing fragmentation.

Together, the Administration's FY2014 Budget and this Strategic Plan will allow us to better achieve a number of inter-related goals:

- It will help Federal STEM efforts reach more students and more teachers more effectively by reorienting Federal policy to meet the needs of those who are delivering STEM education: school districts, States, and colleges, and universities;
- It will help in reorganizing efforts and redirecting resources around more clearly defined priorities, with accountable lead agencies;
- It will enable rigorous evaluation and evidence-building strategies for Federal STEM-education programs;
- It will increase the impact of Federal investments in important areas such as graduate education by expanding resources for a more limited number of programs, while recognizing shortages in key disciplines and professions; and,
- It will provide additional resources to meet specific national goals, such as preparing and recruiting 100,000 high-quality K-12 STEM teachers, recognizing and rewarding excellence in STEM instruction, strengthening the infrastructure for supporting STEM instruction and engagement, increasing the number of undergraduates with a STEM degree by one million over the next decade, and broadening participation in STEM fields by underrepresented groups.

Congressional leadership will be critical to our continued and collective effort on securing the Nation's position at the forefront of STEM innovation and discovery. I look forward to working with legislative leaders on this critical priority.

Sincerely,
John P. Holdren
Assistant to the President for Science & Technology
Director, Office of Science and Technology Policy

About the Office of Science and Technology Policy

The Office of Science and Technology Policy (OSTP) advises the President on the effects of science and technology on domestic and international affairs. The office serves as a source of scientific and technological analysis and judgment for the President with respect to major policies, plans, and programs of the Federal Government. OSTP leads an interagency effort to develop and implement sound science and technology policies and budgets. The office works with the private sector to ensure Federal investments in science and technology contribute to economic prosperity, environmental quality, and national security. For more information, visit http://www.ostp.gov.

About the National Science and Technology Council

The National Science and Technology Council (NSTC) was established by Executive Order on November 23, 1993. This Cabinet-level council is the principal body within the executive branch that coordinates science and technology policy across the diverse entities that make up the Federal research and development enterprise. Chaired by the President, the membership of the NSTC consists of the Vice President, the Director of the Office of Science and Technology Policy, Cabinet Secretaries and Agency Heads with significant science and technology responsibilities, and other White House officials.

The NSTC is organized into five primary committees: Science, Technology, Engineering, and Mathematics (STEM) Education; Science; Technology; Environment, Natural Resources and Sustainability; and Homeland and National Security. Each of these committees oversees subgroups focused on different aspects of science and technology. One of the NSTC's primary objectives is to establish clear national goals for Federal science and technology investments in an array of areas that span virtually all the mission areas of the executive branch. The Council prepares coordinated interagency research and development strategies to form investment packages that are aimed at achieving multiple national goals.

For additional information concerning the work of the National Science and Technology Council, please visit the NSTC website.

About the Committee on Science, Technology, Engineering, and Mathematics (STEM) Education

The NSTC Committee on STEM Education (CoSTEM) coordinates Federal programs and activities in support of STEM education pursuant to the requirements of Sec. 101 of the America COMPETES Reauthorization Act (2010).[2] CoSTEM addresses education and workforce policy issues; research and development efforts that focus on STEM education at the PreK-12, undergraduate, graduate, and lifelong learning levels; and current and projected STEM workforce needs, trends, and issues. CoSTEM performs three functions: review and assessment of Federal STEM education activities and programs; with the Office of Management and Budget, coordination of STEM education activities and programs across Federal agencies, and; development and implementation of a Federal STEM education 5-year Strategic Plan through the participating agencies to be updated every 5 years.

About the Federal STEM Education 5-Year Strategic Plan

The America COMPETES Reauthorization Act (2010)[3] directs OSTP to create an interagency committee under the NSTC to develop a Federal STEM education 5-year strategic plan.

In response to the Act, CoSTEM chartered the Federal Coordination in STEM Education (FC-STEM) Task Force to develop the *Federal STEM Education 5-Year Strategic Plan* and to prepare this report with CoSTEM oversight. The FC-STEM includes members from the 12 Federal agencies represented on CoSTEM, and the Smithsonian Institution. Representatives from each of the agencies represented on CoSTEM made significant contributions to the plan.[4]

CoSTEM chartered the NSTC Fast-Track Action Committee on Federal Investments in STEM (FISTEM) to design and conduct an inventory of current Federal investments in STEM education. FI-STEM included members from the 12 Federal agencies represented on CoSTEM.[5] The *Federal Science, Technology, Engineering and Mathematics (STEM) Education Portfolio*[6], released in December 2011, provided the most-detailed overview of Federal STEM education activities to date, and guided the development of the Strategic Plan. The portfolio report summarizing the FY 2011 investments is included in this document as Appendix A.

MEMBERS OF THE FEDERAL COORDINATION IN STEM EDUCATION TASK FORCE (FC-STEM)

Co-chairs

Joan Ferrini-Mundy (NSF)

Leland Melvin (NASA)

Committee Members

Department of Agriculture
Ann Bartuska
Department of Commerce
Louisa Koch (NOAA)
Department of Defense
Laura Adolfie Department of Education Camsie McAdams Department of Energy Michelle Fox Julie Carruthers

Department of Health and Human Services
Bruce Fuchs (NIH) Department of Homeland Security Stephanie Willett Department of the Interior Marcia McNutt (through 2/13)
Department of Transportation
Timothy Klein

Environmental Protection Agency
Brandon Jones Executive Office of the President
Kumar Garg (OSTP)
Michael Feder (OSTP) (through 2/13) National Aeronautics and Space Administration Leland Melvin National Science Foundation
Joan Ferrini-Mundy

MEMBERS OF THE COMMITTEE ON STEM EDUCATION (COSTEM)

Co-chairs

John P. Holdren (OSTP)

Cora Marrett (NSF, from 3/13)

Subra Suresh (NSF, through 3/13)

Executive Secretary
Carolyn Knowles (NASA)

Committee Members

Department of Agriculture
Catherine Woteki
Department of Commerce
Kathy Sullivan

Department of Health and Human Services Caya Lewis
Department of Homeland Security Daniel Gerstein

Executive Office of the President John P. Holdren (OSTP) Tom Kalil (OSTP)
Steve Robinson (DPC)

Department of Defense
Reginald Brothers
Department of Education
Jim Shelton

Tony Miller (through 4/13) Department of Energy Patricia Dehmer

William Brinkman (through 4/13)

Department of the Interior Anne Castle Department of Transportation

Gregory Winfree Environmental Protection Agency James H. Johnson, Jr.

Mary Cassell (OMB)
National Aeronautics and Space Administration
Leland Melvin National Science Foundation
Cora Marrett (NSF, from 3/13) Subra Suresh (through 3/13) Smithsonian Institution
Claudine Brown

DEPARTMENT AND AGENCY ABBREVIATIONS

Department of Agriculture	USDA
Department of Commerce	DOC
Department of Defense	DOD
Department of Education	ED
Department of Energy	DOE
Department of Health and Human Services	HHS
Department of Homeland Security	DHS
Department of the Interior	DOI
Department of Transportation	DOT
Environmental Protection Agency	EPA
National Aeronautics and Space Administration	NASA
National Institute of Standards and Technology (part of DOC)	NIST
National Institutes of Health (part of HHS)	NIH
National Oceanic and Atmospheric Administration (part of DOC)	NOAA
National Science Foundation	NSF
National Science and Technology Council	NSTC
Nuclear Regulatory Commission	NRC
Office of Management and Budget	OMB
Office of Science and Technology Policy	OSTP
Smithsonian Institution	SI
United States Geological Survey (part of DOI)	USGS

1. EXECUTIVE SUMMARY

"One of the things that I've been focused on as President is how we create an all-hands-on-deck approach to science, technology, engineering, and math. We need to make this a priority to train an army of new teachers in these subject areas, and to make sure that all of us as a country are lifting up these subjects for the respect that they deserve."

President Barack Obama
2013 White House Science Fair; April 2013

Advances in science, technology, engineering, and mathematics (STEM) have long been central to our Nation's ability to manufacture better and smarter products, improve health care, develop cleaner and more efficient domestic energy sources, preserve the environment, safeguard national security, and grow the economy. For the United States to maintain its preeminent position in the world it will be essential that the Nation continues to lead in STEM, but evidence indicates that current educational pathways are not leading to a sufficiently large and well-trained STEM workforce to achieve this goal. Nor is the U.S. education system cultivating a culture of STEM necessary for a STEM-literate public. Thus it is essential that the United States enhance U.S. students' engagement in STEM disciplines and inspire and equip many more students to excel in STEM.

Investing in STEM education is critical to the Nation and its economic future for a number of reasons:

- The jobs of the future are STEM jobs: The demand for professionals in STEM fields[7] is projected to outpace the supply of trained workers and professionals. Additionally, STEM competencies are increasingly required for workers both within and outside specific STEM occupations. A recent report by the President's Council of Advisors on Science and Technology (PCAST) estimates there will be one million fewer STEM graduates over the next decade than U.S. industries will need. [8]
- Our K-12 system is "middle of the pack" in international comparisons: Among 33 Organization for Economic Cooperation and Development (OECD) countries that participated in a recent Programme for International Student Assessment (PISA) study, which measures students' ability to apply what they have learned in reading, mathematics, and science and has been designed to assess whether students can use their knowledge in real-life situations[1], 12 countries had higher scores than did the United States in science and 17 had higher scores in mathematics.[9]
- Progress on STEM is critical to building a just and inclusive society: STEM participation and achievement statistics are especially disturbing for women and minorities, who are substantially underrepresented in STEM fields. While earning a STEM degree is one important milestone in pursuing a STEM career, just 2.2 percent of Hispanics and Latinos, 2.7 percent of African Americans, and 3.3 percent of Native Americans and Alaska Natives have earned a first university degree in the natural sciences or engineering by age 24.[10] While women constitute the majority of students on college campuses and roughly 46 percent of the workforce, they represent less than one in five bachelor's recipients in fields like computer science and engineering, and hold only 25 percent of STEM jobs.[11]

A Strategic Plan for Federal Investment in STEM Education

Many of the CoSTEM agencies have placed a high priority on STEM education and have developed education initiatives unique to their agency's mission, needs, and resources. To better leverage these assets and expertise, the Administration is releasing this STEM education strategic plan, the result of extensive cross-agency collaboration, to articulate a

[1] http://www.oecd.org/pisa/aboutpisa/

strategy for making progress on this national priority. The Administration, through the CoSTEM agencies, is committed to laying groundwork that will set the course for a coherent and impactful collective Federal STEM education investment for the next five years.

The Plan begins by providing an overview of the importance of STEM education to American scientific discovery and innovation, the need to better prepare students for today's jobs and those of the future, and the importance of a STEM-literate society (section 2) and also describes the current state of Federal STEM education efforts (section 3). The document then presents five priority STEM education investment areas where a coordinated Federal strategy can be developed, over five years, designed to lead to major improvements in key areas. This increased coordination is intended to lead to maximum impact and, as it is implemented, will lead to strategies for closer and more effective coordination among agencies with STEM investments (section 4).

Also included in this plan are initial implementation roadmaps in each of the priority STEM education investment areas, proposing potential short-, medium-, and long-term objectives and strategies that might help Federal agencies achieve the outlined goals (section 5). Additionally, throughout the document, the plan highlights (1) key outcomes for the Nation and ways Federal agencies can contribute, (2) areas where agencies will play a lead role, thereby increasing accountability, (3) methods to build and share evidence, and (4) approaches for decreasing fragmentation.

Choosing National Goals That Federal Agencies Can Contribute to

The STEM Strategic Plan sets out ambitious national goals to drive Federal investment in five[12] priority STEM education investment areas:

- *Improve STEM Instruction:* Prepare 100,000 excellent new K-12 STEM teachers by 2020, and support the existing STEM teacher workforce;
- *Increase and Sustain Youth and Public Engagement in STEM:* Support a 50 percent increase in the number of U.S. youth who have an authentic STEM experience each year prior to completing high school;
- *Enhance STEM Experience of Undergraduate Students:* Graduate one million additional students with degrees in STEM fields over the next 10 years;
- *Better Serve Groups Historically Under-represented in STEM Fields:* Increase the number of students from groups that have been underrepresented in STEM fields that graduate with STEM degrees in the next 10 years and improve women's participation in areas of STEM where they are significantly underrepresented; and,
- *Design Graduate Education for Tomorrow's STEM Workforce:* Provide graduate-trained STEM professionals with basic and applied research expertise, options to acquire specialized skills in areas of national importance, mission-critical workforce needs for the CoSTEM agencies, and ancillary skills needed for success in a broad range of careers.

STEM Education Coordination Approaches

Central to the success of this Strategic Plan is moving toward a new approach to coordinating Federal investments in STEM education (Section 4.2). By designating initial lead and collaborating agencies in some of the priority STEM education investment areas, the Strategic Plan encourages a more deliberative focus among new and existing efforts, the expansion of existing collaborations, and the creation of new synergies. The intent is to establish a coordinated, coherent portfolio of STEM education investments across the Federal Government so efforts and assets are deployed effectively and efficiently, for greatest potential impact. To do so, Federal agencies will focus on two main STEM education coordination approaches:

- *Build new models for leveraging assets and expertise.* Implement a strategy of lead and collaborating agencies to leverage capabilities across agencies to achieve the most significant impact of Federal STEM education investments.
- *Build and use evidence-based approaches.* Conduct STEM education research and evaluation to build evidence about promising practices and program effectiveness, to be used across agencies, and share with the public to improve the impact of the Federal STEM education investment.

Congressional leadership and commitment to STEM education stimulated the call for this plan and has been critical in its development. The main body of this report further describes the priority STEM education investment areas and STEM education coordination approaches, and provides initial implementation roadmaps to achieve these strategic objectives. The Administration, including the COSTEM agencies, looks forward to working with legislative leaders on its continued refinement and implementation.

2. INTRODUCTION

> We want to make sure that we are exciting young people around math and science and technology and computer science. We don't want our kids just to be consumers of the amazing things that science generates; we want them to be producers as well. And we want to make sure that those who historically have not participated in the sciences as robustly — girls, members of minority groups here in this country — that they are encouraged as well. We've got to make sure that we're training great calculus and biology teachers, and encouraging students to keep up with their physics and chemistry classes.... It means teaching proper research methods and encouraging young people to challenge accepted knowledge."
>
> **President Barack Obama**
> National Academy of Sciences
> April 2013

Increasing opportunities for young Americans to gain strong STEM skills is essential if the United States is to continue its remarkable record of success in science and innovation. Numerous advances, from mapping the human genome to discovering water on Mars to developing the Internet, would not have been possible without a skilled and creative STEM workforce. New technologies and STEM knowledge lie at the core of our ability to manufacture better, smarter products, improve health care, preserve the environment, and safeguard national security. Individuals prepared with the skills and knowledge to invent, build, install, and operate those new technologies are essential. In addition, a basic understanding of STEM topics and concepts is necessary beyond the workplace in order for citizens to make informed decisions on issues that are increasingly at the center of local and national political debates, such as environmental regulation. STEM literacy is also critical when it comes to making sound personal consumer choices, from health-care decisions to purchases at the grocery store.

STEM knowledge and skills are in even greater demand as the United States confronts a fiercely competitive international marketplace where the advantage goes to companies that are the first to invent and produce innovative products. From 2000 to 2010, the growth in STEM jobs was three times greater than that of non-STEM jobs.[13] The Department of Commerce estimates that in the coming years STEM occupations will grow 1.7 times faster than non-STEM occupations.[14] Furthermore, Georgetown University's Center on Education and the Workforce projects that America will create 779,000 jobs between 2008 and 2018 that require a graduate degree in a STEM field but, based on current trends, only 550,000 native-born Americans will earn STEM graduate degrees during this period.[15]

Industry reports of inability to fill large numbers of STEM-related jobs and other workforce data raise concerns about U.S. capacity to meet the growing demand for STEM workers with appropriate training and skills.[16] The private and public sectors, including the Federal agencies, rely on the U.S. education system and outside-of-school learning opportunities to equip youth with fundamental knowledge and skills needed for choosing and training for their careers, and for making informed decisions as workers, parents, and members of the public.

Furthermore, recent reports indicate that:

- Students who report early expectations for a career in science are much more likely to complete a college degree in a STEM field than students without those expectations. This suggests that early exposure to science topics, at middle grades or below, may be important for a student's future career aspirations.[17]
- The achievement gap in mathematics and science remains a persistent issue. According to one recent report on international assessment of mathematics and science, the science scores of white U.S. eighth graders were surpassed only by the scores of three counties (Singapore, Chinese Taipei, and Korea), while Hispanic and black U.S. eighth graders had scores equivalent to those of students in countries ranked in the bottom third of the 45 countries that participated in the 8th grade science assessment.[18] Only one in five high school graduates who scores in the top quartile in mathematics goes on to become a STEM professional.[19]
- Fewer than 40 percent of students who enter college intending to major in a STEM field complete a STEM degree.[20]
- Only 19 percent of U.S. bachelor's degrees are awarded in STEM fields, while in China over 50 percent of first degrees are awarded in STEM fields.[21]
- Underrepresented minorities in STEM now account for almost 40 percent of K-12 students in the U.S.; however, they earn only 27 percent of all associate's degrees

from community colleges, 17 percent of the bachelor's degrees in the natural sciences and engineering, and 6.6 percent of the doctorates in those fields.[22,23]

- Roughly 30 percent of chemistry and physics teachers in public high schools did not major in these fields and have not earned a certificate to teach those subjects.[24]
- Women make up nearly 50 percent of the U.S. workforce and a majority of college students, but hold less than 25 percent of STEM jobs[25] and earn less than one in five bachelor's degrees in high growth fields like computer science and engineering.[26]

Inadequacies in education pathways leading to STEM degrees and into the workforce amplify concerns that the United States is failing to keep pace with its international competitors in producing a workforce with the necessary skills and knowledge to advance STEM fields.

Given that accomplishing the complex missions of many of the CoSTEM agencies necessarily relies upon a strong and skilled STEM talent pool, those agencies have placed a high priority on trying to improve STEM education and workforce training and developing education and preparation initiatives unique to their respective missions, needs, and resources. *The Federal Science, Technology, Engineering, and Mathematics Education 5-year Strategic Plan* detailed here is intended to improve the efficiency, coordination, and impact of these federally supported STEM education investments.[27]

The Strategic Plan proposes approaches for Federal agencies to coordinate these activities and leverage these assets to make progress on the national priority of STEM education. To do so, the Strategic Plan identifies five priority STEM education investment areas and two coordination strategies for organizing Federal investments in STEM education. Through execution of the Strategic Plan, Federal agencies will collectively address critical STEM challenges with the intention of improving the impact of Federal STEM investments and making faster progress toward an ambitious vision. The Plan builds upon steps the Federal agencies have already taken to improve their STEM education plans and evaluation strategies and to coordinate their efforts28 29 30 and takes the next step toward realizing the shared commitments of the Administration and the Congress to improve STEM education.

3. Federal Role in STEM Education

The 14 Federal CoSTEM agencies invest in and support programs and activities to improve STEM education. These agencies also maintain and fund such important assets as laboratories, research instruments, and facilities, and employ engaged and knowledgeable scientists, researchers, and engineers. Many of their programs are designed to develop a STEM-literate population and to ensure a highly qualified workforce in agency-related fields as well as in the STEM fields more generally.

To do so, they support all levels of learners and all learning environments, including preschool, K-12, two- and four-year colleges, universities, and informal-learning environments. Many programs and investments are also designed to provide learning resources for the general public, including publications, websites, television programs, museum exhibits, after-school programs, and video resources, among others. Federal support for STEM education currently enables:

- Preparation and professional development of STEM teachers and undergraduate faculty in their subject areas, in pedagogy, and in instructional practice through pre-service education and continuing professional development, as well as efforts to enhance teacher recruitment and retention;
- development of instructional materials, learning resources, and courses, including materials that can be integrated into curricula (such as videos, assignment and activity ideas, computer visualizations and simulations), and platforms for building and delivering interactive online courses and learning objects;
- training and re-training to match U.S. workforce skills to the demands of a rapidly-changing global economy and the STEM workforce to accomplish the missions of Federal agencies;
- direct support to students in disciplines related to agency missions through scholarships, fellowships, immersion research experiences in agency operations, training grants, internships, and other programs;
- research and development to understand and improve STEM education and learning programs at all levels, including research on: STEM learning and instructional strategies, learning in informal environments, ways to improve the preparation and professional development of teachers and faculty, STEM workforce development educational programs, and broadening participation education efforts;
- availability of facilities and staff to institutions engaged in STEM education;
- data collection initiatives and program evaluations; and
- public education and lifelong learning projects, including publications, websites, videos, simulations, television programs, museum exhibits, and public events.

Individual agency support of STEM education has evolved as a result of a combination of factors, including the mission and goals of the agency, statutory roles and responsibilities,[2] Congressional and Presidential directives to engage in particular aspects of STEM education, and the unique assets each agency has to contribute.

Most of the CoSTEM agencies support STEM education to meet their specific missions and workforce needs. Some also leverage their facilities, assets, technical workforce, and expertise to support fundamental STEM education research and development and to develop and implement programs that promote STEM literacy or proficiency in mission-related areas. The Federal agencies, including but not limited to the National Oceanic and Atmospheric Administration and the Environmental Protection Agency, that are responsible for maintaining natural resources and environments and for protecting life and property also support STEM education by making their spaces available to educational efforts. This is one of a number of strategies that Federal agencies use to improve stewardship of natural resources and informed decision making by the general public. Agency stewardship programs also typically include outreach, communication, and community engagement activities.

The majority of the Federal investment in STEM education – which totals approximately $3 billion annually (see Appendix A) – is funded largely through programs executed at the NSF, the Department of Education (ED), and, in the biomedical sciences, at the National

[2] Several Federal agencies have statutory responsibility to provide STEM education. These include the Department of Commerce [DOC], Department of Energy [DOE], the Department of the Interior [DOI], Department of Defense [DOD], the National Aeronautics and Space Administration [NASA], and the U.S. Department of Agriculture [USDA].

Institutes of Health. NSF, within its mission to support basic science research and STEM improve education, invests in advancing knowledge through research and development of tested models about how to understand and improve STEM learning. This encompasses teaching and learning from the K-16 through post-graduate levels, as well as continuing education, retraining, and informal education in out-of-school settings. NSF also has a well-established graduate fellowship program that spans all areas of STEM except the biomedical sciences. NSF also has a history of partnering with a number of other CoSTEM agencies in areas of science and education.

ED supports programs to improve education in the United States and has a broad mission to promote student achievement and preparation for global competitiveness. Although only a small fraction of ED's funding supports specific STEM education programs, STEM initiatives have been a competitive priority in such significant programs as Race to the Top[31] and the Investing in Innovation Fund.[32] Given its mission, ED does not have substantial direct access to science and engineering research activities or to a STEM workforce through substantial in-house or external research and development (R&D) programs, but is developing approaches to partner effectively with the other CoSTEM agencies. ED brings unmatched reach to schools, teachers, and students across the Nation, and so this plan provides approaches for leveraging these important connections. Furthermore, ED is building a staff with expertise in STEM teaching, and the National Center for Education Statistics cultivates data critical for STEM education research, including for efforts like the Institute of Education Sciences (IES) Mathematics and Science Education Research Grants Program.

The Smithsonian Institution (the Smithsonian), founded in 1846 and comprising19 museums and galleries, the National Zoological Park, and nine research facilities, promotes a number of engagement opportunities across a variety of disciplines. In 2012, the Smithsonian welcomed 30 million visitors to its public institutions and 103 million users to its website. In the context of Federal STEM activities, the Smithsonian has unique capabilities for serving as a clearinghouse that can reach a wide public audience.

Particular agencies, because of their mandates and goals, have specific expertise and assets to bring to the Federal STEM education investment in collaboration with the designated lead agencies. NASA has exciting assets associated with space exploration, and the USDA has networks that reach across the country through 4-H, extension services programs, and regional networks and laboratories. The NIH oversees traineeship and fellowship programs with a focus on graduate and post-doctoral research that prepare tomorrow's leading biomedical scientists. In addition, agencies such as NASA, HHS, DOD, DOC, DOE, and USDA support the teaching of mission-related science and engineering through connections to the scientific and technical assets that they oversee. Many of the CoSTEM agencies invest in or manage scientific, technological, engineering, and mathematics experts, research facilities and technology, data sets, and natural resources that can be leveraged as sites for STEM learning and workforce training. Some agencies (e.g., NSF, ED/Institute for Education Sciences) focus on building and gathering evidence about effective professional development practices or models, while others (e.g. ED) have mechanisms for scaling up effective practices or systemic changes in professional development.

This diversity of missions and approaches has over time led to an uncoordinated Federal investment in STEM education and STEM programs have proliferated to the point where in FY 2012, there were 226 programs across 13 different agencies described in Appendix A.

This distributed approach to making critical investments in STEM education has made it difficult to ensure that Federal efforts are coherent, strategic, and leveraged for greatest impact. At the same time, the activities supported by the agencies have important functions and with coordination the combined efforts can unquestionably be greater.

Aside from STEM-specific education programs, the Federal Government also supports the development of STEM skills through investments that have goals that extend beyond STEM education (e.g., the many general education programs at ED), or those that focus on STEM research. At the postsecondary level, a portion of the Federal research and development (R&D) expenditure supports investigator-driven research awards at colleges and universities where research teams may include STEM-focused undergraduate students and graduate research assistants. While science R&D funding provides critical training opportunities for aspiring STEM graduates, and contributes to the development of a skilled workforce, such investments are outside the scope of STEM education considered in this Plan. Although the Federal Government plays an important role in STEM education, it cannot achieve success by itself. To effectively leverage its investments, the Federal Government must coordinate its efforts strategically and collaborate with non-Federal partners to support institutional, state, and local efforts. Local and state education agencies, institutions of higher education, professional and scientific societies, philanthropic and corporate foundations, aquaria, botanical gardens, museums, science centers, after-school providers, and private industry, for example, play potentially significant roles in growing our Nation's STEM education pipeline and creating pathways to STEM. Each stakeholder brings a set of resources and expertise that are necessary for our Nation's STEM education systems to reach their full potential. The Federal Government works closely with these stakeholders to identify common areas of concern and to collaborate on strategies that leverage Federal and non-Federal assets, and this Strategic Plan creates an opportunity for more of these key partnerships. Efforts such as those described in BOX 1 illustrate the potential additive power of partnerships that amplify the contributions of the Federal Government and other stakeholders.

Box 1. *100kin10* Coalition

Building on a recommendation of the President's Council of Advisors on Science and Technology (PCAST), President Obama in his 2011 State of the Union address called for a new effort to prepare 100,000 excellent STEM teachers over the next decade with strong teaching skills and deep content knowledge.

In response to that call to action, more than 150 organizations to date have joined together in a coalition called *100Kin10* to make over 100 measurable commitments to increasing the supply of excellent STEM teachers; hiring, developing, and retaining excellent STEM teachers; and building the 100Kin10 movement. With leadership from Carnegie Corporation of New York, the coalition and its partners have raised over $52.5 million from a broad range of foundations and philanthropists under a unique "funding marketplace" model where funders can choose from among a registry of high-quality proposals. The 100Kin10 coalition has also become a network with great potential for disseminating ideas, with leading-practice case studies, capacity-building workshops, and funds available for competitive research and collaboration grants.

4. STRATEGIC PRIORITIES AND COORDINATION STRATEGIES

This Strategic Plan provides a shared roadmap to ensure that the collective Federal investment of 14 agencies in STEM education has the best potential for substantial impact in key priority areas. In addition, the Strategic Plan has been developed after carefully considering the findings of CoSTEM *Portfolio* reports analyzing Federal investments in STEM. (See Appendix A for the *FY 2011 Federal Science, Technology, Engineering and Mathematics (STEM) Education Portfolio Highlights.)*

Strategic vision: A future where:

- The United States has a well-qualified and increasingly diverse STEM workforce able to lead innovation in STEM-related industries and to fulfill CoSTEM agency workforce needs;
- American students have access to excellent P-12, postsecondary, and informal STEM education and learning opportunities; and
- Federal STEM education programs are based on evidence and coordinated for maximum impact in priority areas.

Achieving this vision will require strong CoSTEM agency commitment to prioritization and coordination. The remainder of this plan sets forth five priority STEM education strategic investment areas and two goals for STEM education coordination.

4.1. Priority STEM Education Investment Areas

The priority areas identified by CoSTEM were selected on the basis that enhanced and coordinated Federal investments in these areas seem especially likely to accelerate progress toward the strategic vision outlined above. These areas reflect the needs of our Nation, align with priorities of both the Administration and Congress, and can draw effectively on Federal STEM-related assets. Through this Plan and Federal agencies' implementation efforts, the intention is to achieve significant, measurable impacts in five priority STEM education investment areas: (1) improve P-12 STEM instruction; (2) increase and sustain youth and public engagement in STEM; (3) improve undergraduate STEM education; (4) better serve groups historically underrepresented in STEM fields; and (5) design graduate education for today's STEM workforce.

Over the course of the implementation of the Strategic Plan, Federal agencies will strive to align their STEM education investments in support of those five priority investment areas. Through this effort, existing collaborations will be strengthened, and new collaborations will be undertaken. It is likely that additional areas of importance will emerge over time, and as progress is made on these priorities they may be adapted and focused. Also, it is important to note that these national goals are meant to provide overall direction and focus, and that Federal agencies will need to work together to achieve them. Progress on how Federal agencies are contributing will be assessed by developing metrics and milestones for the strategic objectives, introduced in section 5.

1. Improve STEM Instruction. Prepare 100,000 Excellent New K-12 STEM Teachers by 2020, and Support the Existing STEM Teacher Workforce

Research shows that top-performing teachers make a dramatic difference in student achievement and suggests that for students who learn from these teachers year after year, achievement gaps narrow significantly.[33, 34, 35, 36, 37, 38] Every STEM student deserves an excellent teacher – one whose preparation has included appropriate opportunities to learn his or her subject and to participate in clinical experiences in schools. Every teacher and education leader deserves access to the preparation, on-going support, recognition, and collaboration opportunities needed for success. However, not all teachers receive the support they need, and science and mathematics teachers are among the most difficult to recruit and retain in K-12 schools.[39, 40]

To increase the number of excellent K-12 STEM teachers, CoSTEM agencies will undertake increased coordination among STEM-teacher preparation, professional development, support, and recognition efforts within existing and proposed programs. Continued research on teacher learning and STEM teacher development is essential to guiding these implementation and coordination efforts. Appropriate connections to local and state policy, standards, and assessments will be encouraged to ensure that the Federal investments are designed for local impact and implementation.

2. Increase and Sustain Youth and Public Engagement in STEM. Support a 50 Percent Increase in the Number of U.S. Youth Who Have an Effective, Authentic STEM Experience Each Year Prior to Completing High School

Progress for the Nation in STEM depends not only on a STEM workforce, but on a public that understands the role of STEM in addressing societal issues and is prepared to use STEM knowledge in personal and professional settings. For many Americans, both students and adults, opportunities for STEM learning occur through effective engagement across classroom and out-of-school settings, as each year tens of millions—of all ages and backgrounds — engage with science in informal settings ranging from pre-school programs to museums, to national parks, as well as cyberspace, all of which provide potential venues for advancing science literacy. This engagement [41] is critical to the learning process and to selection and persistence in STEM careers. Through effective engagement, learners have the chance to develop interest in and positive attitudes toward STEM topics, as well as improved perception of their ability to participate in STEM, all of which can lead to improved understanding and proficiency.[42, 43] Exciting trends today related to STEM engagement include digital badges that encourage students to earn progressive recognition in the form of online and mobile distinctions for acquiring cumulative skills, the Maker Movement, low-cost tools that allow students to design and make just about anything, and "games for learning." Additionally, citizen science activities can advance both science and learning.

By "authentic STEM experience," CoSTEM means a designed experience inside or outside of school in which learners engage directly in doing STEM. This broad designation covers a range of commonly referenced notions, from "hands-on" science, to problem-based learning, to inquiry. Part of the challenge in implementing steps toward this goal will be defining and studying whether experiences that are purported to be "authentic" have measurable impact on student motivation, persistence, and learning.

Agency implementation of this priority area will involve the development of coordinated programs inside and outside of school, platforms, and infrastructure to provide desired audiences with STEM experiences enabled through government assets, including but not limited to materials, facilities, and skilled STEM professionals. CoSTEM agencies - many of which employ skilled scientists and engineers, develop advanced instruments and facilities, and conduct research in STEM - currently play a key role in both formal and informal education engagement efforts. CoSTEM agencies also generate a wealth of information and data electronically available to the public, which can enable exploration of science or engineering subjects as part of classroom learning to encourage general interest, or as citizen science.

Collaboration among all agencies will be critical for better understanding the current baseline of engagement activities and their outcomes and furthering a focused and impactful government investment in engagement.

3. Enhance STEM Experience of Undergraduate Students. Graduate One Million Additional Students with Degrees in STEM Fields Over the Next 10 Years

Poor retention rates among undergraduate STEM majors in U.S. institutions remain a major concern. Only 43 percent of students entering as a STEM major in a four-year public college or university graduate with a STEM degree. Worse, only about 14 percent of community college students who declare a STEM major on entry are still in a STEM field at the time of their last enrollment.[44]

Economic projections suggest the need for as many as one million additional STEM professionals over the next decade above current graduation rates.45, 46, 47 That includes STEM majors at a variety of skill and knowledge levels from community college graduates, to quality STEM teachers, to scientists and engineers with advanced degrees. Outside of STEM careers, other fields increasingly require employees to have a good foundation in STEM disciplines.

In addition, a scientifically and computationally literate citizenry is increasingly necessary to critically evaluate personal and societal issues that increasingly rely on science and technology underpinnings.

Through internship, scholarship, and fellowship programs, research experience opportunities, initiatives to create and test innovative instructional approaches and materials, faculty professional development, and research on STEM learning, Federal agencies can create strong emphasis on improving undergraduate STEM education. In addition, efforts to broaden participation in STEM will be critical to reaching the number of needed graduates, and the resulting diversity will enhance innovation in STEM fields.

4. Better Serve Groups Historically Underrepresented in STEM Fields. Increase the Number of Underrepresented Minorities That Graduate College with STEM Degrees in the Next 10 Years and Improve Women's Participation in Areas of STEM Where They Are Significantly Underrepresented

Groups underrepresented in STEM fields include Hispanics and Latinos, African Americans, American Indians, Alaska Natives, Native Hawaiians and Pacific Islanders, the economically disadvantaged, people with disabilities, and women and girls. According to the most recent census, underrepresented minority groups (URGs) make up approximately 28 percent of the U.S. population but are experiencing the greatest population growth—with an 11 percent increase for African Americans and a 37 percent boost for Hispanics between 2000 and 2009.[48] Minority populations are expected to increase to 54 percent of the U.S. population by 205049, [50] and student diversity is reflected in this demographic change. Although the student population is becoming increasingly diverse, URGs remain underrepresented in STEM education and careers. Moreover, women, who comprise 48 percent of the U.S. workforce, have remained below parity and make up only 24 percent of STEM professionals. Underrepresentation of these groups in STEM fields begins early and persists across the P-12, post-secondary, and STEM-workforce spectrum.

To make progress in this priority area, CoSTEM agencies will work with relevant stakeholder communities, including faculty, administrators, and students from Minority-Serving Institutions (MSIs) given that approximately 10 percent of the funds invested by Federal agencies on this priority are focused on these institutions. Agencies will consider emphasizing education at critical transition points from P-12 to postsecondary education and from postsecondary education to the STEM workforce, when students from groups traditionally underrepresented in STEM often drop out of the STEM pipeline. As part of this focus, agencies will work to create more common definitions and consistent categorization of programs that serve underrepresented groups as either a focus or an emphasis. This may also include improving access to and increasing coordination across, programs for Minority-Serving Institutions.

5. Design Graduate Education for Tomorrow's STEM Workforce. Provide Graduate-Trained STEM Professionals with Basic and Applied Research Expertise, Options to Acquire Specialized Skills in Areas of National Importance and Mission Agency's Needs, and Ancillary Skills Needed for Success in a Broad Range of Careers

The National Center for Science and Engineering Statistics projects that between 2010 and 2020, 2.6 million jobs will require an advanced degree.[51] Filling these positions will be critical to the Nation's economic prosperity and global competitiveness. Recent reports by the Council of Graduate Schools, the NIH, the National Research Council, and professional societies[52] have called for shifts in approaches to STEM graduate education. The Federal Government supports graduate students primarily through research assistantships on individual investigator awards, which generally involves student participation in the research activities of the principal investigator. Mechanisms such as fellowships and traineeships also provide significant opportunities for graduate students, and increasingly there is a need to prepare students not only for academic research positions, but also for career options in the private and government sectors.

As an initial focus under this strategic investment priority area, CoSTEM agencies will coordinate to improve access to, and efficacy of, government-funded graduate fellowships. Over time, CoSTEM agencies may also consider addressing improvements to a broader range of Federal approaches to graduate student support.

Federally funded graduate fellowships recognize students with high potential in STEM research and innovation. These fellowships provide an opportunity for students to pursue unique educational opportunities and research across disciplinary boundaries, often with significant autonomy. Importantly, fellowships are designed to provide support directly to individual students to pursue their research interests. Thus, seeking a better way to leverage and coordinate fellowships provides a promising starting point in promoting and encouraging new opportunities to prepare the science and engineering workforce of tomorrow. There is growing evidence that opportunities for professional development of students to learn a broader range of skills that are important in STEM fields (e.g., communication) and to participate in applied work on authentic problems and challenges of government and the private sector are important components of graduate education.

Furthermore, tomorrow's STEM workforce will need to include effective change makers and entrepreneurs in business, public service, civil society, and academia. Some universities are encouraging students to set and meet more ambitious goals for their research, education, and service; giving students greater autonomy earlier in their career; connecting students to real-world problems at a regional, national, and global level; and involving students in the design of university curricula, research initiatives, and collaborations with external partners. The CoSTEM agencies can continue to benefit from understanding the effectiveness of innovative practices that universities are incorporating in their graduate education to promote those practices in graduate education more broadly.

4.2. STEM Education Coordination Approaches

Coordination of Federal investments in STEM education, including mission-agency STEM-workforce training and education, is central to the success of this Strategic Plan is. The two coordination approaches are (1) building new models for leveraging assets and expertise; and (2) identifying, using, and sharing evidence-based approaches. The intent is to establish a coordinated, coherent portfolio of STEM education investments across the Federal Government so efforts and assets are deployed effectively and efficiently, for greatest potential impact. By designating initial lead and collaborating agencies within certain priority areas, the Strategic Plan will encourage a more deliberate focus among new and existing efforts and create new synergies among programs and agencies. Likewise, many agencies already are committed to the use of evidence and rigorous evaluation in budget, management, and policy decisions, and to building the internal capacity to do so. That can also be supported and enhanced in a number of ways, including by creating and using common metrics, evidence guidelines, and evaluation practices; by developing complementary program goals; and by creating a shared understanding of evidence-based STEM education practices.

1. Build New Models for Leveraging Assets and Expertise. Implement a Concept of Lead and Collaborating Agencies in Priority Areas to Leverage Capabilities Across Agencies to Ensure the Most Significant Impact of Federal STEM Education Investments

Efficient and effective use of funding and resources remains a priority for Federal investments, including those in STEM education. Efficiency across agencies can be created through inter-agency coordination and collaboration. Efficiency within agencies will occur through alignment of investments with capabilities, roles, and missions. An emphasis on investments designed to be complementary, that leverage available resources, will lead to a strategic portfolio intended for maximum impact.

To achieve this leveraging, lead agencies have been identified in several priority areas for this Strategic Plan. All agencies will employ a range of strategies to bring coherence and efficiency into STEM education programs with similar goals, including aligning programmatic goals, building common infrastructure, developing joint solicitations or memoranda of understanding, consolidating programs, and using new funding strategies, such as performance partnerships that encourage agency collaborations.[53]

Lead agencies will be responsible for convening other CoSTEM agencies, for helping to facilitate review and revision with collaborating agencies, and for tracking progress toward achieving Administration priorities. In each priority area, all CoSTEM agencies will be invited to participate. The following have been designated as the initial lead agencies for CoSTEM implementation:

- *Department of Education (ED),* which will assume a convening role to initiate planning around the priority area of "Improve STEM Instruction." The early focus of these discussions will be around how to best mobilize Federal assets and resources toward the improvement of STEM teacher preparation and continuing teacher professional development; how to connect these efforts with state and local policies and contexts; and how to leverage partnerships for improvement.
- *National Science Foundation (NSF),* which will play a convening role around the priority area of "Enhance STEM Experience of Undergraduate Students," including assisting with improving the delivery of undergraduate STEM education through evidence-based reforms. NSF will also work alongside NIH, USDA and other CoSTEM agencies on improving graduate fellowships. Efforts will include sharing detailed information about agency and disciplinary STEM workforce needs at the post-baccalaureate and postdoctoral levels, and mapping existing Federal resources and assets as a basis for designing strategic collaborations and synergies.
- *Smithsonian Institution,* which will focus its convening efforts on the priority goal of "Increase and Sustain Youth and Public Engagement in STEM." As part of the effort, Smithsonian will work with NSF, ED, the other CoSTEM agencies including NASA, NOAA, USDA, NIH, DOI, and other science partners best understand the agencies' unique expertise and resources for engaging learners with science, and will explore existing and potential approaches to improving infrastructure and access.

Designation as a lead agency does not narrowly define an agency's role in STEM education. It does not mean that NSF, for instance, will abandon its efforts in areas other than the improvement of undergraduate STEM education, or that no other agency will have any role in undergraduate education. It means instead, that NSF may request additional funding and resources to support an increased role initially in leading the improvement of undergraduate STEM education. Similarly, while ED is the initial lead for P-12 instruction, it will also play a strong role in supporting engagement activities and building bridges between in-school and out-of school learning to increase the effectiveness of both. The other CoSTEM agencies will be key collaborators, working with the lead agencies to find ways to build on their existing investments in STEM education, and leverage the passion and expertise of their staff and other STEM professionals who will continue to provide access to STEM content and Federal assets that can be used in formal and informal learning environments.

Following on the work that will be done within the initial convenings around the priority areas, CoSTEM will create implementation subcommittees for the STEM education investment priority areas and interagency coordination approaches, as appropriate. The Implementation Subcommittees would be responsible for tasks including:

- Reviewing roadmaps and implementation plans prepared by initial lead and collaborating agencies;
- Tracking implementation of the education investment priority areas within the Strategic Plan;
- Developing and monitoring metrics for progress, assessing whether efficiencies and improved impact are achieved, and recommending adjustments to the implementation process; and,
- Developing a framework and process for more coordinated Federal STEM education budget planning.

2. Build and Use Evidence-Based Approaches. Conduct STEM Education Research and Evaluation to Build Evidence About Promising Practices and Program Effectiveness, Use Across Agencies, and Share with the Public to Improve the Impact of the Federal STEM Education Investment

There is a strong interest across the government in using evidence in policy and management decisions in all arenas, including STEM education, to increase the impact and efficiency of Federal programs. Given the important role of data and evaluations, Federal STEM education investments will increase use of research studies and program evaluation to identify and test practices and approaches for their effectiveness, thus building the evidence base for improving Federal STEM education investments. As appropriate, programs will rest on or scale up evidence-based practices. Findings about effectiveness will be shared, along with research-based understandings, and where indications of success are strong, agencies will be encouraged to act upon them. Where evidence is only suggestive of promise, but there are other rationales for making investments of a particular type, programs and investments will be designed so that an evidence base is built as the program is implemented.

Where appropriate, CoSTEM Implementation Subcommittees, together with lead and collaborating agency staff, will determine the most suitable approaches to synthesize evidence-based practices and processes for accumulating data, evaluations, information, and metrics and will consider collaborative evaluation approaches within and across agencies.

CoSTEM has developed a set of design principles (see Appendix B) that represent initial frameworks for Federal investments to be successful based on current evidence and best practices. As appropriate, these will be updated over time as new research and findings emerge. In addition, the impact and utility of information produced by STEM education investments will be considered in evidence-based grant-making strategies (e.g., pay for success and tiered-evidence grants). There is also work that is available to begin to identify such practices within the What Works Clearinghouse (http://ies.ed.gov/ncee/wwc), including reviews of studies of interventions and practice guides that identify opportunities for additional development and validation work.

5. Implementing the Federal STEM Education 5-Year Strategic Plan

For each of the five priority STEM education investment areas, a set of strategic objectives has been identified as the focus for initial implementation of the Strategic Plan. These objectives have been developed with several considerations: they strive to be specific, so that progress and impact can be measured; they are meant to align with the strengths and assets of the designated lead agency and to allow for the significant role of collaborating agencies; and they represent areas where there is a clear responsibility for involvement of the Federal Government, with the realization that Federal investment will play only a part in achieving the intended impact.

Implementation of the Strategic Plan will require commitment of the Administration, agency leadership, and legislative leaders in allocating the resources for the collaboration, coordination, and evaluation that will be necessary to realize these goals. Implementation also will require oversight and the development of more detailed roadmaps by lead and collaborating agencies over the coming months. CoSTEM will assist in this process.

5.1. Implementation of STEM Education Priority Investment Areas

In order to make progress on the priority STEM education investment areas, lead and collaborating agencies will determine which features of current Federal investments could be adapted, modified, or leveraged in support of the coordinated activities of the lead agencies, and how. This collaborative process will ultimately determine how agencies can best create infrastructure in priority areas and how significant strengths and assets at collaborating agencies can be made available in conjunction with these efforts. This will require coordination of agency priorities and goals, statutory requirements, authorizing legislation, and available expertise and resources. Close collaboration among lead and collaborating agencies will be needed to maximize the opportunities to continue to meet any mission-specific needs.

Preliminary implementation roadmaps are provided below for each of the five priority STEM education investment areas that, where appropriate, may help guide future budget planning and requests, mechanisms for investment, communication and outreach with stakeholder communities, and reassessment of evaluation plans and practices within each of the priority areas. The ability of agencies to implement the strategic objectives and to create common metrics will depend on agency capacity and will require financial commitments to ensure adequate capacity both at lead agencies, for design and oversight of programs, and at collaborating agencies, for asset identification and coordination functions. The implementation of the Strategic Plan will include a process for periodic examination of overall progress on all strategic objectives, with the expectation that some may be revised, and others will be added.

Implementation Roadmap: Improve STEM Instruction

IMPACT STATEMENT: Prepare 100,000 Excellent New K-12 STEM Teachers by 2020, and Support the Existing STEM Teacher Workforce

The education and development of teachers of STEM involves their pre-service preparation and also continuing professional development and learning opportunities over the course of their careers. The retention of teachers includes strategies that can involve school- and district-based resources and professional supports, meaningful community-based initiatives, and exciting opportunities to learn STEM and participate in authentic STEM research. Excellent STEM teachers know how to provide effective instruction and to inspire interest in these subjects so that their students understand STEM concepts and skills, and the Federal Government has a strategic role in providing learning and growth opportunities for the STEM teaching workforce.

Background

Of all of the activities that occur within the formal and informal education systems, the interactions among teachers, learners, and the content is the primary determinant of student success in grades K-12. Research shows that top-performing teachers can make dramatic differences in student achievement and suggests that the impact of assigning students to top-performing teachers each year can significantly narrow achievement gaps.[54, 55, 56, 57] Thus, the need to focus on improving STEM teaching is clear.

The scale of K-12 STEM education is large. There are more than 3.6 million full-time teachers in the Nation's elementary and secondary schools.[58] Of these, about 500,000 teach at least one course in either mathematics or science at the middle- or high-school levels.[59] Nationwide, there are about 2,800 universities and colleges that prepare teachers. These institutions are accredited by states or national organizations to certify teachers. Additionally, there are "alternative" programs to prepare teachers. According to the National Center for Alternative Certification, there are 136 state-defined alternate routes to teacher certification and about 600 alternate route programs that are implementing these options.[60] Policies for teacher credentialing and continuing licensure are developed at the state level, as are standards for K-12 STEM curricula.

The need for more STEM teachers is well documented,[61] but the recruitment and retention of STEM teachers is challenging. States routinely list a lack of teachers in mathematics and the sciences as one of their most pressing labor shortages.[62] The pre-service preparation of STEM teachers, activities that occur largely within individual institutions of higher education and are governed by state policies, does not compare favorably to the preparation of teachers in countries whose students show high levels of achievement in STEM.[63] Additionally, there is continued debate on the specific STEM content that teachers need to be effective.[64,65] The emerging field of discipline-based educational research (DBER) documents the difficulties of teaching undergraduate STEM courses for understanding and, along with PCAST's *Engage to Excel* report, promotes widespread use of evidence-based instructional practices – including in undergraduate courses where future teachers are learning STEM content.[66] And, as technologies that can support and improve classroom learning are readily available, teachers need to have opportunities to learn and use these exciting and powerful tools.

Once teachers complete their initial preparation, local districts and schools plan and pay for the majority of their professional development activities, and much of this professional development is focused on general classroom and pedagogical issues rather than subject-specific topics. Federal investment in teacher professional development is, conservatively, only 9% of the total Federal STEM investment (Appendix A). The quality of this professional development ranges considerably, and research is limited about its impact on teachers' effectiveness in enabling student learning. [67, 68, 69] In addition, CoSTEM agencies often offer a wide range of learning opportunities intended to inspire STEM teachers. In light of recent findings that access to learning opportunities and school supports are critical in keeping STEM teachers in classrooms,[70] it is important that the professional development made available to teachers is connected to their day-to-day work, related to state standards (including the Common Core State Standards in Mathematics and the Next Generation Science Standards) and to the ongoing professional credentials that are required in states.

Federal investments across lead and collaborating agencies alone cannot possibly reach all STEM teachers, or ensure that the offerings provided are compatible with the policies and practices of the states, districts, and schools where teachers are located. Thus, the focus for the Federal investments in teacher education, both pre-service and in-service, needs to be on building well-tested, replicable models that can be scaled or adapted across teacher education institutions, and at the level of states and districts for implementation, and providing infrastructure that can allow for scaling. These models can then be aligned with local policies and processes in ways that support teachers in their schools and classrooms. In addition, developing partnerships with local entities, local and regional science facilities that can support high quality STEM content, and the private sector will be especially critical.

In his 2011 State of the Union address, President Obama called for a new effort to prepare 100,000 STEM teachers over the next decade with strong teaching skills and deep content knowledge. The President's call built on key conclusions of the President's Council of Advisors on Science and Technology (PCAST) - that teachers need to have enough content knowledge to link STEM to compelling real-world issues, model the process of scientific investigation, effectively address student misconceptions, and help their students learn to reason and solve problems like mathematicians, scientists and engineers.

Significant Related Efforts

The *FY 2011 Portfolio* report indicates that, for FY 2011, about 10 percent ($315 million) of Federal STEM Education funding was dedicated to investments reporting to have a primary objective of supporting pre-service and in-service STEM educators. Federal agencies funded an additional $925 million in investments that include improvement of STEM teacher education as a secondary goal. Of the investments that primarily focus on improving teacher education, 78 percent supported the professional development of in-service educators, and the remainder supported both pre-service and in-service educators. The majority of the dollars came from the NSF and ED, through large programs such as ED and NSF's Mathematics and Science Partnership (MSP) programs, Teacher Loan Forgiveness, and the Robert Noyce Scholarship Program.

Strategic Objectives

ED will serve initially as the lead agency for the goal of improving K-12 STEM instruction. All other CoSTEM agencies will serve as collaborating agencies. Collectively, Federal agencies will focus investments that would contribute to achieving the President's goal of preparing 100,000 new excellent STEM teachers, with strategies such as:

- Identify, develop, test, and support effective teacher preparation efforts that encourage teachers' use of evidence-based practices that provide students with rich STEM learning opportunities; and
- Increase the number and quality of authentic STEM experiences[71] for pre- and in-service P12 teachers participating in federally supported internship, fellowship, and scholarship programs

Table of potential actions/outcomes/metrics

Identify, develop, test, and support effective teacher preparation efforts that encourage teachers' use of evidence-based practices that provide students with rich STEM learning opportunities		
Actions	**Outcomes**	**Metrics/Milestones**
Near-term (years 1-2)		
Identify and assess Federal investments that incentivize (a) the recruitment or training of excellent K-12 STEM teachers to work in high-need schools and (b) the retention of those effective STEM teachers whose students show annual growth in STEM learning.	Better understanding of Federal investments that incentivize (a) the recruitment or training of excellent P-12 STEM teachers to high-need schools and (b) the retention of those effective STEM teachers whose students show annual growth in STEM learning.	Federally funded STEM teacher education initiatives that: (a) provide data on recruitment or training of excellent STEM teachers in high-need schools; (b) provide data on the retention of effective STEM teachers in high-need schools; (c) include data about teacher education practices, both pre-service and in-service education and experiences; (d) relate participation in these programs to student outcomes; and (e) as applicable, hold state educational agencies and local educational agencies accountable for implementing evidence-based practices that lead to (1) recruitment of excellent P-12 STEM teachers to high-need schools and (2) retention of effective teachers whose students show annual growth in STEM learning.

Actions	Outcomes	Metrics/Milestones
Mid-term (years 3-4)		
Use the evidence base amassed to develop guidance for effective Federal programs that incentivize (a) the recruitment or training of excellent P-12 STEM teachers to high-need schools and (b) the retention of those effective STEM teachers whose students show annual growth in STEM learning.	Collecting research findings about the nature of Federal programs that incentivize (a) the recruitment or training of excellent P-12 STEM teachers to high-need schools and (b) the retention of those effective STEM teachers whose students show annual growth in STEM learning. As appropriate, Federal program investments implement guidance to increase the percentage of excellent P-12 STEM teachers (a) recruited or trained to teach in high-need schools and (b) retained when their students show annual growth in STEM learning.	Federally funded STEM teacher education programs including pre-service and in-service education and experiences, that: (1) have incorporated feedback to monitor the use of evidence-based practices and (2) monitor the evidence to be aware of changes that could impact programs.
Long-term (4+ years):		
Federal funding initiatives are designed to include: (1) recruitment and retention incentives; (2) evidence-based practices; and (3) the collection of necessary data to document the impact of teachers' participation in the Federal programs on (a) the increased percentage of excellent P-12 STEM teachers recruited to high-need schools, and (b) the increased percentage of effective STEM teachers retained whose students show annual growth in STEM learning.	Regular updates on the status of Federally-funded teacher programs that incentivize (a) the recruitment or training of excellent P-12 STEM teachers to high-need schools and (b) the retention of those effective STEM teachers whose students show annual growth in STEM learning. Maintain the advances obtained and institutionalize accountability for program outcomes.	Federally funded STEM teacher education programs in each agency, which offer pre-service and in-service education and experiences focused on recruitment and retention, will monitor the quality of the teacher education programs involved and assess the impact of the investment on improving (a) the recruitment of excellent P-12 STEM teachers to high-need schools and (b) the retention of those teachers whose students show annual growth in STEM learning.

Increase the number and quality of authentic STEM experiences[72] for pre- and in-service P-12 teachers participating in Federally supported internship, fellowship, and scholarship (IFS) programs.		
Actions	**Outcomes**	**Metrics/Milestones**
Near-term (years 1-2)		
Identify and assess Federal STEM investments in teacher IFS, including: size, scope, structure, methods, status of assessment and evaluation activities and characteristics of included STEM experiences. Initiate plans for linking existing resources with new infrastructures for reaching wide audiences.	Collect information about Federal investments in teacher IFS including recommended additional information beyond that which was available. A baseline of relevant programs will provide the basis upon which assessment criteria can be developed. These criteria will be used to compare/contrast practices to develop guidance.	Agencies with teacher IFS that: (1) provide participant data on IFS; (2) include data about teacher education practices; (3) hold organizations accountable for implementing evidence-based practices; and (4) are capable of connecting participation in authentic STEM experiences with desired pupil outcomes.
Mid-term (years 3-4)		
Use the evidence base to develop guidance for effective program design of scholarships to support teacher candidates.	Investments implement guidance to increase effectiveness.	Agencies with teacher IFS that: (1) have incorporated feedback to monitor the use of evidence-based practices and (2) monitor the evidence to be aware of changes that could impact programs.
Long-term:		
Federal IFS funding initiatives are designed to include: (1) evidence-based practices; (2) grade-appropriate authentic STEM experiences; and (3) collection of necessary data to document the connection between teachers' participation and desired student outcomes.	Biennial report on the status of Federally-funded teacher IFS programs. Maintain the advances obtained and institutionalize accountability for program outcomes.	All Federally funded teacher IFS programs monitor the quality of the teacher education programs involved.

Implementation Roadmap: Increase and Sustain Youth and Public Engagement in STEM

IMPACT STATEMENT: Support a 50 Percent Increase in the Number of U.S. Youth Who Have an Effective, Authentic STEM Experience Each Year Prior to Completing High School

Engagement is part of the larger process of learning. It is the critical component for capturing the learner's interest and involvement, and for inspiring further development of knowledge and understanding. To ensure a STEM-skilled workforce of the future, the Federal Government has considerable assets that can engage youth so that the pathway through their education leads to the challenging STEM-related careers of tomorrow, and to a "culture of STEM" in the public.

Background

Given the importance of STEM engagement at a national level, a common understanding of the terminology is necessary. As defined in *The 2011 Federal STEM Education Portfolio*, investments that focus on engagement are designed to increase learners' involvement and interest in STEM, inform their view of STEM's value in their lives, or positively influence the perception of their ability to participate in STEM. The scope of STEM engagement is vast, and includes investments in a wide range of areas, such as development of learning materials; programs at museums, science centers, or parks; games, simulations, and virtual environments; "Citizen Science" initiatives; public talks, and educational broadcast programming. Thus, there are many avenues available to the Federal Government for reaching learners, and many possibilities for tailoring content for STEM-learning audiences.

Research indicates that instructional approaches or learning opportunities that engage students actively increase skill acquisition and information retention, encourage more positive attitudes toward STEM disciplines, and strengthen retention of students in STEM majors.[73, 74] For example, learning theory and empirical evidence about how people learn[75] suggest that STEM experiences that engage learners in "active learning" improve retention of information and critical thinking skills.[76] Furthermore, research studies in STEM education support this positive relationship between STEM engagement experiences and student achievement. For example, one study demonstrated that a 2- to 3-week university Summer Science Academy program, which provided students in grades 6-12 with field trips, laboratory experimentation, and other authentic, science-related, learning opportunities nonexistent in the students' schools, produced significant, long-term outcomes on student achievement and attitudes toward science.[77] This growing body of research literature on effective STEM engagement and evidence-based practices can guide Federal Government investments.

Additionally, agencies are already exploring or supporting a number of promising new and emerging strategies in the area of youth STEM engagement and have started to generate great momentum within grassroots networks and crowd-driven virtual platforms. For instance, the Maker Education Initiative, which grew out of a network of self–directed STEM enthusiasts who share their creations with students and the general public through Maker Faires across the world, recently announced its intention to create a Maker Corps of student maker leaders. These youth, embedded in informal learning institutions across the country, will provide hands-on learning opportunities to interested students, exciting their desire to be, as the President has said, "makers of things, not just consumers." Additional new opportunities for engagement include the digital badge movement bolstered by NASA[78] as well as the MacArthur and Mozilla Foundations. Additionally, crowd-sourced collection of data through "citizen science" activities like "Backyard Biofuels," provide an opportunity to evaluate and scale exciting new avenues for inspiring students in STEM fields while building an evidence base for novel activities.

In the years ahead, the Smithsonian Institution will convene agencies to help improve the reach of informal STEM learning opportunities by ensuring that materials are aligned with what students are learning in the classroom as appropriate, and that exciting, emerging science is accessible to students through innovative means. Priority will be placed on effective, authentic STEM experiences. The Smithsonian will work with the CoSTEM agencies such as NASA, USDA, NIH, DOI, NOAA, and other science partners to harness their unique expertise and resources to enable the use of promising materials and curricula, on-line resources, and effective delivery and dissemination mechanisms by more learners both inside and outside the classroom.

Significant Related Efforts

Over half of the Federal Government's STEM education investments identified in *The 2011 Federal STEM Education Portfolio* provided engagement as the primary or secondary objective (156 of 233 programs). Funding for investments with the primary objective of engagement totaled \$164 million. Of these investments, about half (approximately 44 percent) are categorized as large-scale programs. The budget of each investment was under \$30 million.

CoSTEM agencies are heavily involved in engagement. For example, USDA houses the Cooperative Extension System through land-grant universities and includes a mission to bring research, education and extension, and applied research-based practices, to both youth and adults. This occurs through 4-H, a unique public-private partnership with USDA, extension land-grant universities and local governmental entities providing public support along with a private, non-profit national partner, the National 4-H Council. Together with private donors and foundations, these entities support the programmatic efforts for over 6 million youth participants and half a million volunteers, both youth and adults. Since 2010, 4-H's Science Leadership Academies have engaged 1,266 4-H professionals in training workshops with the ultimate goal of equipping more 4-H staff and volunteers with the ability to engage students in informal learning environments.

The Corporation for National and Community Service (CNCS) is launching a new STEM AmeriCorps, multi-year initiative to place hundreds of AmeriCorps members in nonprofits across the country to mobilize STEM professionals with the goal of inspiring young people to excel in STEM education.[79] The U.S. Department of Education also hosts a network of 21st Century Community Learning Centers to provide academic enrichment opportunity during non-school hours for children in high-poverty and low-performing schools. The Smithsonian sponsors a Youth Engagement through Science (YES!) effort, which connects local youth with Smithsonian collections, experts, and training to inspire them to pursue STEM careers.[80]

The Federal Government has a diverse range of assets to increase interest in STEM. These assets include, but are not limited to scientific data, technology, research and engineering facilities, natural environments, science-technology centers, engineers, technologists, and scientists. The agencies use these assets to provide online, place-based (e.g., research and engineering facilities, federally managed lands and waterways, and museums or visitor centers), and other experiential or hands-on learning opportunities primarily in informal learning environments for people of all ages. In addition, NSF provides development and research grants in the area of informal STEM education, while some CoSTEM agencies have legislative mandates to promote direct public participation in exploration, STEM research, and environmental stewardship. The private sector can help in these efforts, leveraging the passion and expertise of its science and technology trained workforce. For example, leading U.S. technology companies have launched an effort called US2020, an all-hands-on-deck effort to have many more STEM professionals mentor children from kindergarten through college.[81]

Alignment, coordination, and continued research on such experiences can support the development of multiple pathways to continued and deeper exploration of STEM learning in informal learning environments, connections between formal and informal learning settings,[82] and the impact of engagement experiences on outcomes such as student learning. Together, these efforts will move the STEM engagement enterprise forward while building evidence-based practices. In this rich and diverse context for providing engagement, the Smithsonian will assume an initial lead role for coordination and developing the infrastructure to allow the more efficient and open access[83] to these extraordinary Federal resources.

A key focus within the Engagement priority is on assessing the impact of engagement experiences with appropriate outcome measures and better understanding the connections between inspiration and excitement and persistence and achievement in STEM for school and careers.

Strategies

With the Smithsonian playing an initial lead coordination role, Federal agencies will focus investments on three main strategies.

- Federal investments in engagement that draw explicitly on the scientific, technological, and engineering assets (e.g., facilities, scientific and engineering staff, instruments, data, and federally-managed public lands and waterways) of the Federal Government, where feasible, to provide authentic experiences;
- Federal engagement investments that support the integration of STEM into existing school readiness and after-school programs with significant local, regional, or national reach; and
- Federal engagement investments that contribute to improving empirical understanding of how engagement in authentic STEM experiences relates to improved student learning or interest outcomes.

Table of potential actions/outcomes/metrics

Federal investments in engagement that draw explicitly on the scientific and engineering assets (e.g., facilities, scientific and engineering staff, instruments, data, and Federally managed public lands and waterways) of the Federal Government to provide authentic experiences.		
Actions	Outcomes	Metrics/Milestones
Near-term (years 1-2)		
Identify scientific and engineering assets that are being effectively leveraged in existing investments on engagement. Build additional infrastructure to provide access to Federal engagement assets.	Identify agencies and assets that have evidence-based models of best practices in engagement.	Number of collaborations that draw on STEM assets with evidence of impact for engagement.
Mid-term (years 3-4)		
Establish new approaches to make mission agency assets available through new networks and other agency collaborations to wider groups of youth.	Identification of opportunities to form connections between engagement assets and programs.	Number of youth that have significant interactions with Federal science agency assets.
Long-term:		
Revise solicitations beyond Federal science mission agencies to tie education research and development to the assets of the mission agencies.	A research and development infrastructure to assess and track impact of engagement experiences.	Increase instruments and assessment techniques that provide valid, reliable assessment of engagement and identification of relevant outcome measures.

(Continued)

Federal engagement investments that support the integration of STEM into existing school readiness and after-school programs with national reach.		
Actions	Outcomes	Metrics/Milestones
Near-term (years 1-2)		
Identify and begin collaborations among a set of programs for school readiness and after school with national reach, including exploration of possibilities with Head Start, 21st Century Community Learning Centers, and 4-H.	Convening to identify model programs and mechanisms to facilitate collaborations.	Number of projects that are being leveraged for national reach.
Mid-term (years 3-4)		
Launch partnerships among agencies and programs to develop curricula, professional development, & implementation.	Development of engagement-related materials and models for partnerships are supported by agencies.	Number of projects that have links to programs with national reach.
Long-term:		
Initiate longitudinal study to examine the relationship between introducing STEM engagement experiences into national programs and student outcomes.	Report on research findings about relationship between authentic STEM experiences and engagement-related outcomes in national programs.	Number of learners reached through national programs that have embedded evidence-based STEM engagement experiences.

Federal engagement investments that contribute to improving empirical understanding of how engagement in authentic STEM experiences relates to improved student learning or interest outcomes.		
Actions	Outcomes	Metrics
Near-term (years 1-2)		
Determine which investments are positioned to collect data on authentic STEM experiences and how they affect student learning, interest, involvement, and/or motivation (engagement) outcomes. Build knowledge base related to youth engagement and criteria to record number of authentic STEM experiences achieved through investments."	Identify availability and quality of data and methodologies to gather data on authentic STEM experiences.	Number of projects from their investments that are participating in the building of the knowledge base of authentic STEM experiences with mechanisms for relating to student engagement.
Mid-term (years 3-4)		
Continue data gathering and analysis efforts, and when appropriate, the development of new metrics and methodologies.	As appropriate, implementation of a data coordination infrastructure to enable research across sites to better understand the relationship between authentic STEM experiences and engagement related outcomes.	Number of projects from their investments that are participating in the building of data coordination infrastructure.

Actions	Outcomes	Metrics
Long-term:		
Encourage Federal engagement initiatives to participate in the data coordination infrastructure.	As appropriate, report on research findings about relationship between authentic STEM experiences and engagement related outcomes.	Have many more Federally funded authentic STEM engagement projects are part of the data coordination infrastructure.

Implementation Roadmap: Enhance STEM Experience of Undergraduate Students

IMPACT STATEMENT: Graduate One Million Additional Students in STEM Fields Over the Next 10 Years

A number of economic and labor analyses suggest that if the United States is to maintain its global preeminence in the STEM fields and benefit from the social, economic, and national security advantages that come with such preeminence, then it must produce approximately one million more STEM professionals than are projected to graduate over the next decade.[84, 85, 86] To meet this goal, the United States institutions of higher education will need to increase the number of students who receive undergraduate STEM degrees by about 34 percent over current rates by 2020. To meet this need, the Federal Government has made preparing one million more STEM graduates the central focus of the Cross Agency Priority (CAP) goal on STEM education, a multi-agency effort to align agency activities and accountability around a clear priorities.[87] The Federal Government, however, is just one of a number of stakeholders needed to reach the goal of graduating one million additional students with STEM degrees, and so attaining it will involve public-private partnerships as well as a Federal focus.

Background

Students with STEM degrees are needed at a variety of skill- and knowledge-levels—from community college graduates trained in information technologies or those involved in the production of advanced manufacturing, materials and energy, to high-quality STEM teachers, to scientists and engineers working in research labs. Besides STEM career areas, other fields increasingly require employees to have a good foundation in STEM disciplines. In addition, it is increasingly important to prepare a scientifically and computationally literate public able to critically evaluate societal issues as diverse as climate change, application of medical technologies, or alternative energy sources. Also in the competitive global work environment there are a number of other competencies that are considered useful[88] such as curiosity, creativity, tolerance of ambiguity, resilience in the face of setbacks, and abilities to work effectively with people who have different perspectives, priorities, or intellectual approaches. Experiences within the STEM education system may provide opportunities for building these competencies.

Poor retention rates in undergraduate STEM majors remain a major concern and a critical factor in achieving the one million more graduates goal. Only 43 percent of students entering a STEM major in a four-year public college or university graduate with a STEM degree. Worse, about 14 percent of community college students who declare a STEM major on entry are still in that field at the time of their last enrollment.[89] PCAST concluded in its recent report, *Engage to Excel: Producing One Million Additional College Graduates with Degrees in Science, Technology, Engineering, and Mathematics*, that retaining more STEM majors is the lowest-cost, most efficient policy option to provide the STEM professionals that the Nation needs.

As noted in the PCAST report, as well as in the Federal CAP goal, increasing the retention of STEM majors to 50 percent would generate approximately three-quarters of the targeted one million additional STEM graduates over the next decade by increasing the annual number of students with bachelor or associate degrees in STEM fields by approximately 75,000. Furthermore, such an increase appears feasible.

Research has uncovered a variety of evidence-based best practices to recruit, engage, and retain STEM majors.[90] These practices include pedagogies, curricula, instruction materials, peer or mentors and other academic and cultural supports, resources, and tools to engage students in the classroom and support their learning. Also, there is evidence that co-curricular activities—such as learning communities and summer "bridge" programs—that provide academic support, and community experiences that encourage students to adopt STEM majors, can have a positive impact on student retention. [91] Application of such practices at both two-and four-year schools, coupled articulation agreements that ensure compatibility of both discipline content and pedagogical practice, are promising strategies to increase significantly the number and diversity of talented STEM majors entering the workforce and graduate education.

BOX 2. Department of Defense (DoD) and NSF: Awards to Stimulate and Support Undergraduate Research Experiences (ASSURE)

Research experiences for undergraduates are important authentic STEM experiences that provide a means for students to gain first-hand appreciation of the knowledge-production process. The Awards to Stimulate and Support Undergraduate Research Experiences (ASSURE) program is a collaboration between DoD and NSF to increase student engagement in STEM. DoD's ASSURE is executed through NSF's Research Experiences for Undergraduates (REU) program to support undergraduate research in DoD-relevant disciplines at NSF-sponsored college and university locations. ASSURE seeks to increase the number of high-quality undergraduate science and engineering majors who ultimately decide to pursue advanced degrees in those fields.

With a well-defined common focus, students gain access to faculty and/or other research mentors and relevant facilities to complete research projects. The research projects can be based in a single discipline or academic department, or in interdisciplinary or multi-department research opportunities with a strong STEM focus, providing a range of possible research experiences for undergraduate students.

Significant Related Efforts

The Federal Government has a number of investments that can be leveraged to improve undergraduate STEM education. Some investments already include collaboration across agencies (See BOX 2). Historically, the majority of funding from investments that at least partially target undergraduate students has the primary goal of graduating students with STEM degrees. The other primary objectives, with at least $100 million dedicated to support undergraduate students, were in the categories of education R&D, institutional capacity building, learning, and STEM careers. An additional substantial portion of funding has been dedicated to programs that provide fellowships, scholarships, and/or internships. In addition, many agencies have programs in place to offer research experiences to undergraduates.

Furthermore, students should be empowered to play a more active role in their own educations through student-led innovation. For instance, the MIT Clean Energy Prize is a student-led, student-focused clean-energy innovation and entrepreneurship prize that has helped train hundreds of students from more than 60 universities around the country with the skills necessary to be clean-energy entrepreneurs. More than 30 companies have been launched as a result of the competition since the prize was founded in 2008.[92] Student-led innovation programs currently allow agencies to directly encourage this community of 20 million technically capable, creative, and energetic innovators to use their skills to work on real-world problems. Engaged and empowered students can increase the ability of universities to organize research, coursework, and experiential learning and to become more engaged universities.

In addition, as outlined in the CAP goal on one million more STEM graduates, the private sector is also taking leadership in addressing challenges of undergraduate STEM education.[93] For example, through such initiatives as the National Undergraduate STEM Partnership, leading universities, industry groups, the Business Higher Education Forum, the Association of American Universities, the Association of Public and Land-Grant Universities, the American Council on Education, the National Defense Industrial Association, and others have committed to improving undergraduate STEM education, particularly in the first two years of college. Private industry also has sought ways to partner with the Federal Government to encourage increased enrollment and retention in STEM fields. For example, through its Graduate 10K+ Initiative, the NSF has partnered with Intel and GE in making grants to institutions whose projects aim to improve retention of undergraduates in engineering and computer science. This effort is funded with $10 million in private-sector donations.[94]

Strategic Objectives

To contribute to the goal of producing one million more STEM graduates in the next 10 years, and building upon the CAP goal in this area, the Federal Government, working closely with partners and with NSF will pursue the following strategies:

- Identify and broaden implementation of evidence-based instructional practices and innovations to improve undergraduate learning and retention in STEM and develop a national architecture to improve empirical understanding of how these changes relate to key student outcomes;
- Improve support of STEM education at 2-year colleges and create bridges between 2- and 4- year postsecondary institutions;
- Support and incentivize the development of university-industry partnerships, and partnerships with federally supported entities, to provide relevant and authentic STEM learning and research experiences for undergraduate students, particularly in their first two years; and
- Address the problem of excessively high failure rates in introductory mathematics courses at the undergraduate level to open pathways to more advanced STEM courses.

In these strategies, agencies will work to incorporate a focus on increasing the participation of women and underrepresented minorities and identifying and supporting innovation in higher education.

Table of potential actions/outcomes/metrics

Identify and broaden implementation of evidence-based instructional practices and innovations to improve undergraduate learning and retention in STEM and develop national architecture to improve empirical understanding of how these changes relate to key student outcomes		
Actions	Outcomes	Metric or Milestone
Near-term (years 1-2)		
As appropriate, establish a cross agency mechanism to obtain data on instructional practices at institutions.	Revisions to solicitations, syntheses of evidence-based practices, and measurement of the use of evidence-based instructional practices introduced in programs, as appropriate.	Number of programs that introduce a focus on evidence-based instructional practices.
Mid-term (years 3-4)		
Agencies partner with stakeholders to develop and implement faculty professional development to use evidence-based practices/design innovation	New partnerships among agencies, professional societies, and higher education to focus on development and use of evidence- based practices.	Improved retention in STEM majors in a selected set of universities participating in partnerships.
Long-term:		
Stakeholders participate in national effort to validate and scale-up evidence on relationship of practices to student outcomes such as retention.	Infrastructure for sharing national data about use of evidence-based practices and relationship to student outcomes has been developed.	National findings available about the relationship between instructional and program characteristics in undergraduate STEM and student outcomes.

Improve support of STEM education at 2-year colleges and create bridges between 2- and 4-year postsecondary institutions.		
Actions	Outcomes	Metric or Milestone
Near-term (years 1-2)		
Gather information about STEM programs and challenges at two-year colleges.	Identification of issues in STEM at two-year colleges.	Better understanding of STEM education at two-year colleges and next steps for agencies
Mid-term (years 3-4)		
Establish network of two-year colleges and 4-year institutions committed to STEM excellence, in partnerships as possible.	Stronger linkages and better articulation agreements between two year and four year colleges and universities and science assets among Federal agencies.	Have many institutions of higher education committed to partnering with Federal agencies to improve 2-to-4 year transitions and STEM opportunities.
Long-term:		
Federal agencies reconfigure or enhance programs as appropriate to support effective partnerships.	Programs and STEM outcomes improve in 2-year colleges.	Number of two-year colleges with effective STEM programs substantially increases.

Support the development of university-industry partnerships, and partnerships with Federally supported entities, to provide relevant and authentic STEM learning and research experiences for undergraduate students, particularly in their first two years.		
Actions	Outcomes	Metric or Milestone
Near-term (years 1-2)		
Identify existing Federal STEM internship and other authentic research opportunities, and assemble available evidence about impact.	A more detailed baseline description of the Federal portfolio.	Federal agencies can better report the number of internships and other STEM research opportunities available to undergraduates, and any available evidence about their impact.
Mid-term (years 3-4)		
Convening of government, higher education, and private sector to share best practices and develop incentives for research and internship experiences for students in the first two years of undergraduate education as called for in the PCAST engage to excel report and the CAP goal.	More commitments from industry and government to increase the availably of internships and other authentic research opportunities for students in the first two years.	Federal programs increase focus on research experiences in the first two years and on assessing impact.
Long-term:		
Federal agencies lead the development of a streamlined application process and potentially establish a database of available research experiences.	Industry and government combine resources for reaching more students.	Increased diversity in the students who participate in undergraduate research experiences in the first two years.

Address the problem of excessively high failure rates in introductory mathematics courses at the undergraduate level, to open pathways to more advanced STEM courses.		
Actions	Outcomes	Metric or Milestone
Near-term (years 1-2)		
As feasible, NSF and ED prepare a joint solicitation to improve K-in 16 mathematics education. As feasible, accounting of mathematics funding in agency portfolios and using that to inform the solicitation.	Joint work between NSF and ED this key area.	Programs that incorporate the expertise or assets of more than one agency to improve passing rates of introductory mathematics courses.
Mid-term (years 3-4)		
Convening on introductory college mathematics, including government and key stakeholders to highlight promising practices, potential partnerships, and key challenges.	Spread of information about effective practices and new partnerships for scaling and identification of challenges and changes	A cohort of colleges and universities that have significant challenges in introductory mathematics performance have initiated changes in programs and practices.

(Continued)

Actions	Outcomes	Metric or Milestone
Long-term:		
National data collection/surveys redesigned to track the transitions of students from high school to undergraduate courses in mathematics.	New databases to allow for monitoring national progress on the mathematics issues.	A cohort of universities and colleges can demonstrate substantial improvement on introductory mathematics performance.

Implementation Roadmap: Better Serve Groups Historically Underrepresented in STEM Fields

IMPACT STATEMENT: Increase the Number of Underrepresented Minorities That Graduate College with STEM Degrees in the Next 10 Years and Improve Women's Participation in Areas of STEM Where They Are Significantly Underrepresented

Members of racial and ethnic minority groups are projected to become the majority of America's population in the next 30 years. Currently, however, they account for only 28 percent of STEM workers.[95] Women make up nearly half of the total workforce, but they constitute only 24 percent of STEM jobholders.[96] To meet growing demands from industry for more STEM-qualified employees continued and improved support for the development of talent from all groups is essential. This will require significant improvements in broadening the participation of underrepresented groups and women in STEM fields.

Background

The underrepresentation of certain groups in STEM fields begins early and persists across the STEM school and workforce spectrum. There are three complementary reasons why participation of underrepresented minorities, lower-income children, and women in STEM is critical to sustaining our capacity to conduct research and innovate: our sources for the future STEM workforce are currently not sufficient; the demographics of our domestic population are shifting dramatically; and diversity of ideas and perspectives in STEM is a strength that benefits both diverse groups and the Nation as a whole.[97, 98] Intentional and consistent efforts are needed along the STEM trajectory in order to increase the number of individuals from groups traditionally underrepresented in STEM that graduate and are well-prepared with STEM degrees, because members of these groups leave STEM majors at higher rates than others. These students deserve special attention and must be a deliberate part of any national strategy because they offer an expanding pool of untapped talents and are a large underutilized source of potential STEM professionals.

While the overall increase in college enrollment is encouraging, the National Academy of Sciences (NAS) report, *Expanding Underrepresented Minority Participation* [87] cited that STEM degree completion rates for underrepresented minorities (URMs) were still lower than that of White and Asian students. Moreover, URMs attending two-year colleges are less likely to major in STEM subjects and earn STEM degrees. In 2009-10, African Americans earned 8.6 percent of all science and engineering bachelor's degrees, including only five percent of all bachelor's degrees in engineering[99] Hispanics earned seven percent of all bachelor's degrees in engineering and computer and information sciences.[100] While women who pursue STEM degrees earn higher proportions of bachelor's degrees in medical and social sciences, they earn lower proportions of bachelor's degrees in engineering and computer and information sciences. In 2009-10, 18 percent of all bachelor's degrees in engineering and computer and information sciences were awarded to women.[101] Data also show that students with disabilities are less likely to major in a STEM field than students without disabilities.[102]

The disparities in access to STEM courses and programs that exist in higher education are also evident at the K-12 level. Data from ED's Office for Civil Rights' Civil Rights Data Collection (CDRC) reveal disparities in access to high-level mathematics and science courses in high school.[103] Notably, of the high schools serving the most Hispanic and African-American students, less than one-third offer calculus, and only 40 percent offer physics. Students with limited English proficiency make up six percent of the high school population, but account for 15 percent of the students for whom algebra is the highest-level mathematics course taken by the final year of their high school career. Only 19 percent of all girls taking at least one Advanced Placement (AP) course were enrolled in AP mathematics compared to 26 percent of all boys taking at least one AP course. Of the African-American or Hispanic students enrolled in AP mathematics, only 39 percent are girls. Girls also represented only 46 percent of the students enrolled in physics. According to CRDC data, students with disabilities represent four percent of students taking Algebra II, chemistry, and physics, 0.1 percent of students taking calculus, and eight percent of students taking biology.

Significant Related Efforts

Promoting equitable participation of people seeking STEM degrees and careers in STEM is a priority of a significant number of Federal programs. In FY 2011, $616 million in STEM education investments identified their primary focus as supporting underrepresented groups. More than 20 Federal STEM education investments, including the 1890 scholars programs at USDA, targeted historically underrepresented groups by supporting Minority-Serving Institutions (MSI), including Historically Black Colleges and Universities (HBCU), Hispanic-Serving Institutions (HSI), Alaska Native-Serving Institutions, Native Hawaiian-Serving Institutions, Insular areas institutions, and Tribal Colleges and Universities. In the past, the largest amount of funding was directed to HSIs, mostly from ED's Developing Hispanic-Serving Institutions STEM and Articulation Program, with additional funds at supporting agencies.

Of the Federal investments that are focused on improving participation of underrepresented groups in STEM (see Appendix A), the majority include support for undergraduate students or institutions (70 percent), while 20 percent include support for K-12 students. Also, for those investments that are focused on broadening participation in STEM, about 5 percent of funding supported research to better understand the science of broadening participation.

Although there has been some progress and limited success in broadening participation of underrepresented groups in STEM, much more can be done to attract and retain these groups. Some of the programs that have shown promise in their efforts in broadening participation include the Educational Partnership with Minority Institutions funded by NOAA (see BOX 3), the LSAMP program at NSF, and non-Federal programs such as the Meyerhoff Scholarship program at University of Maryland Baltimore County, the successful Bayer Scholars program of Duquesne University, and the revamped Computer Science program at

Harvey Mudd College that has increased the percentage of women computer science majors from single digits to 40 percent in just a few years.[104] These programs have provided strong mentoring support, academic development and enhancements, and social and professional networks, and have helped students not only to complete their baccalaureate degrees but also to pursue graduate degrees that strengthen their abilities to obtain STEM jobs.

Moreover, there is a growing body of research demonstrating that teaching students in STEM classes that their intelligence is not a fixed trait, but instead is something that can be developed with effort and help from others (i.e., teaching them a "growth mindset"), can significantly improve academic outcomes, in some cases reducing achievement gaps among minorities and women by 30-50 percent.[105, 106]

BOX 3. NOAA Educational Partnership Program with Minority Serving Institutions

The National Oceanic and Atmospheric Administration (NOAA)'s mission is to understand and predict changes in the Earth's environment and conserve and manage coastal and marine resources to meet our Nation's economic, social, and environmental needs. To maintain a workforce with the scientific skills and technical capacity to meet and advance this mission, the agency requires a diverse pool of candidates that reflect the American population. To that end, in 2001 NOAA established the Educational Partnership Program (EPP) with Minority Serving Institutions (MSI) to: (1) increase the number of post-secondary students, particularly from underrepresented communities, who are educated in STEM disciplines that directly support NOAA's mission; (2) contribute to NOAA's mission by strengthening and building capacity in NOAA science and management areas primarily at MSIs; and, (3) leverage NOAA funds to strengthen the education, training and research capacity at the EPP Cooperative Science Centers (http://epp.noaa.gov).

Each of the four Cooperative Science Centers (CSCs), selected via a competitive grants program, is led by an MSI that collaborates with at least six colleges or universities. In addition to the CSCs, the EPP has a Graduate Sciences Program, an Undergraduate Scholarship Program, and an Environmental Entrepreneurship Program. Since its inception, the program has supported more than 1,100 students obtain undergraduate and graduate degrees in fields related to NOAA's mission; 119 of those students obtained PhDs. To date, 19 percent of science professionals from underrepresented communities who have been hired by NOAA had been supported by EPP.

Additionally, the Federal Government also has engaged in a number of collaborative efforts alongside private and non-profit and academic sector partners with the primary or secondary intention of increasing opportunity and access for students who are typically underrepresented in STEM. For example, the Equal Futures Partnership, a multi-country initiative launched by then-Secretary of State Hillary Clinton and other world leaders in September 2012, features commitments from a diverse set of corporate partners seeking to open doors to high-quality educational and career opportunities for women in STEM disciplines.

Strategic Objectives

In order to contribute to the goal of improving STEM graduation success rates for underrepresented minorities and women, Federal agencies will work to implement three strategies:

- Be more responsive to the rapidly changing demographics and issues for particular groups and particular STEM fields through investments in broadening the participation of groups traditionally underrepresented in STEM;

Be more responsive to the rapidly changing demographics and issues for particular groups and particular STEM fields through investments in broadening the STEM participation of groups traditionally underrepresented in STEM.		
Actions	Outcomes	Metric or Milestone
Near-term (years 1-2)		
Agencies and stakeholders identify specific issues and concerns and encourage new ideas, from communities, and the field. Specifically focus on challenges associated with increased access to AP courses by military children, underrepresented minorities; individuals with disabilities, and on special initiatives to encourage women in computer and physical sciences and engineering. Where possible, increase participation of targeted underrepresented minorities, lower-income children, individuals with disabilities, military children, and women in internships at Federal research and engineering facilities.	Where appropriate, convening with MSIs and other stakeholders, together with Federal agencies, to develop common understandings of key challenges.	Proposed ideas for improving impact of Federal portfolio of investment in broadening participation across agencies.
Mid-term (years 3-4)		
Develop key partnerships among agencies and external organizations to design new courses or redesign existing ones to make them relevant to a growing number of URMs in STEM as well as make them competitive for entrance into college.	Have several agencies partner and develop mechanisms that encourage participation of URMs in taking Advanced Placement and other courses, (including on-line) which will allow them to gain college credits in various STEM courses.	Increased support for more relevant courses for URMs will be enhanced, thereby increasing opportunities for URMs in STEM.
Long-term:		
National databases and systems allow for more detailed monitoring of progress by group and STEM field.	Agencies and professional organizations are able to monitor progress of involvement of particular underrepresented groups in particular fields.	Increased participation of groups nderrepresented in particular STEM fields is evidenced through national data sources.

- Focus investments on developing and testing strategies for improving STEM preparation for higher education for students from groups underrepresented in STEM; and
- Invest in efforts to create campus climates that are effective in improving success for students from underrepresented groups through mentorship, technical assistance, and other innovative practices.

Table of potential actions/outcomes/metrics

Focus investments on developing and testing strategies for improving STEM preparation for higher education for students from groups underrepresented in STEM.		
Actions	Outcomes	Metric or Milestone
Near-term (years 1-2)		
Review current models of supporting MSIs to identify gaps and complementary models in the Federal program portfolio and create logical pathways and linkages.	Where appropriate, conduct studies on the various models used by Federal agencies and design of new programs to fill gaps.	New programs will address gaps in the models and will increase the ability of the agencies to support more students from URMs and women, while complementary programs will offer more opportunities and access to students.
Mid-term (years 3-4)		
Synthesis and analysis of evidence-based practices emerging from research and development of more common metrics to assess the impacts of broadening participation using these analyses.	A set of guidelines or measures will be available to "test-drive" on Federal programs that support broadening participation.	Federal investments are coordinated to build knowledge base about effective strategies, and use common metrics to assess program impacts.
Long-term:		
Agencies will work with academia, industry, and education organizations to identify mechanisms that should be more widely implemented.	Graduation rates in STEM fields in both two-year and other higher education institutions for students from all groups traditionally underrepresented in STEM are increasing in key STEM fields	Clearer understanding of how effective approaches can be adapted for a variety of institutions.

Invest in efforts to create campus climates that are effective in improving success for students from underrepresented groups through mentorship, technical assistance, and other innovative practices.		
Actions	Outcomes	Metric or Milestone
Near-term (years 1-2)		
Coordinate efforts to ensure that students belonging to URGs are not excluded in the campus environment by engaging STEM education experts.	In FY2013, Federal agencies will develop a document that will serve as a guideline for a campus climate conducive to all students, particularly those belonging to URGs and will be disseminated to grantees, stakeholders and other relevant organizations.	Campus activities, policies and practices will reflect awareness of and agreement with the guidelines.

Actions	Outcomes	Metric or Milestone
ED's Office for Civil Rights (OCR) will continue to update a STEM resources webpage it recently created to make clear to schools, students, and parents that the civil rights laws enforced by OCR apply to STEM courses and programs.	Enhanced understanding among stakeholders of civil rights steps needed to improve awareness of and compliance with regard to access to Title IX within STEM programs and contexts.	In FY 2013 will establish a baseline for providing guidance and tools to improve compliance.
Mid-term (years 3-4)		
Federal Government encourages higher education institutions that can be model demonstration sites for effective STEM campus climates.	Increased visibility for institutions that are effective in STEM-friendly campus climates for underrepresented groups, within a year of implementation.	Programs increasingly directly address campus climate.
Long-term:		
Higher education institutions that are not MSIs but that have significant numbers of students from underrepresented groups agree to adapt and implement policies and practices to change campus climate for more effective STEM education.	A gathering of institutions sponsored by professional higher education organizations to highlight the challenges of changes in campus climate.	Increased investment in research in the science of broadening participation as it relates to campus climate issues.

As appropriate, CoSTEM will determine appropriate mechanisms for launching a national conversation among MSI leaders and others to engage the broad stakeholder community in determining appropriate steps for launching the implementation of these strategic objectives, including potentially designating a lead agency for the effort.

Implementation Roadmap: Design Graduate Education for Tomorrow's STEM Workforce

IMPACT STATEMENT: Provide Graduate-Level Trained STEM Professionals with Basic and Applied Research Expertise, Options to Acquire Specialized Skills in Areas of National Importance and Mission Agency's Needs, and Ancillary Skills Needed for Success in a Broad Range of Careers

Background

The preparation of graduate students in science and engineering contributes to global competitiveness, producing the highly skilled workers of the future and supporting the research needed for a knowledge-based economy. It is projected that between 2010 and 2020, 2.6 million jobs will require an advance degree, including non-STEM and professional degrees.[107] Given this need, it is critical that the training of today's scientists and engineers, at all levels, but particularly specialized advanced education, take into account the speed of technological innovation, the nature and practice of science and engineering, innovation as an economic driver, and the challenging issues facing society.

Recent reports by the Council of Graduate Schools (CGS), National Research Council, and the NIH all have made recommendations for how to better prepare graduate students. Their recommendations include strengthening professional development and deepening employer-university engagement. Those recommendations also are reflected in a recent PCAST report[108] that underscores the need to provide incentives for industry to invest in research and university-industry partnerships and to change educational programs to prepare graduates for a broad range of careers. Given that approximately 50 percent of doctorate holders work in fields outside of academia,[109] it is critical that the training and preparation students receive enables them to succeed in the workforce.

According to *Science and Engineering Indicators* 2012,[110] there were about 440,000 full time graduate students in U.S. institutions in 2009. Of these about 81,000 or 18 percent were supported by Federal funds. Of those supported by Federal funds, over 70 percent were on supported as research assistants on research grants, about 10 percent on fellowships, 10 percent on traineeships, and the remaining on Federally funded teaching assistants and other mechanisms. These funding mechanisms are critical in developing the STEM workforce and advancing U.S. research and innovation.

CoSTEM will initially focus on coordination to improve the access to, and efficiency of, graduate fellowships in STEM. In this regard, NSF will coordinate with agencies such as NIH (biomedical), NASA (aerospace), EPA (environmental), etc., as well as other agencies to understand specific needs and support applied research with regard to particular STEM fields and to help ensure that students have access to specialized training in order to prepare the Nation's STEM workforce. Roles and responsibilities may be defined, for instance through Memoranda of Understanding established between NSF and other CoSTEM agencies where appropriate, to help ensure their interests and STEM workforce needs are addressed. Over time, CoSTEM agencies may also consider addressing improvements to a broader range of Federal approaches to graduate education.

Significant Related Efforts

The Graduate Education Modernization (GEM) informal working group, which includes DOE, NIH and NSF, has noted that a critical issue confronting U.S. competitiveness is the alignment of competencies required for success in the broader STEM professional workforce with the preparation provided by graduate education and training. A desired outcome of a broad GEM effort is an expanded collaborative effort between academic institutions and Federal funding agencies to help ensure that the graduate education enterprise provides enhanced opportunities to prepare students for a wide range of career paths, without compromising preparation for the academic research enterprise that currently relies heavily on graduate students to advance knowledge and innovation. For example, NIH expects all recipients of the Ruth L. Kirchstein National Research Service Award institutional research training grants to provide trainees with professional development skills and career guidance.[111] More recently, NIH's the National Institute of General Medical Sciences has identified some useful ways to implement the actions in their Strategic Plan for Biomedical and Behavioral Research Training, such as encouraging exposure to multiple career path options for students and recognizing that well-prepared graduates must have a deep knowledge in a specific field and be conversant in related fields.[112]

The informal Interagency Workgroup for STEM Graduate Fellowships (IWGF), which has included the graduate fellowship directors from all the CoSTEM agencies, seeks to strengthen the ongoing and future diverse pools of highly trained scientists and engineers in the United States. The IWGF goals are to (1) discuss common issues among federally funded graduate fellowship programs and (2) create more efficient mechanisms across all stages of fellowships management ranging from pre-award (the competition and review processes) to post-award (fellow's professional development, graduation rates, etc.).

Strategic Objectives for Federally Funded Fellowships

In order to contribute to the goal of improving coordination on and access to graduate fellowships in STEM, Federal agencies will focus on three strategies:

- Recognize and provide financial support to students of high potential for contributions in science and engineering fields, and for success in their ultimate careers;
- Provide opportunities for fellows' preparation in areas critical to the Nation and, in particular, critical to preparing the workforce needed to advance the missions of Federal agencies, including scholarship-for-service and STEM-focused pathways to Federal service programs throughout the Federal research and engineering enterprise; and
- Continue and enhance mechanisms that evaluate the impact of fellowships to inform future Federal investments.

Table of potential actions/outcomes/metrics

Recognize and provide financial support to students of high potential for contributions in science and engineering fields, and for success in their ultimate careers.		
Actions	Outcomes	Metric or Milestone
Near-term (years 1-2)		
Develop a coordinated Federal approach to fellowships that increases efficiency and effectiveness for agencies and applicants, including "one stop shopping" features.	Implementation of coordinated approach to fellowships that increases efficiency and effectiveness for agencies and applicants.	Increased number of opportunities in areas of national need offered to increased number of fellows.
Mid-term (years 3-4)		
Design of a comprehensive approach to fellowships that increase efficiency and effectiveness for agencies and applicants.	Established process for identifying areas of national need and developing opportunities in those areas.	Process identified with steps for implementation.
Long-term:		
Coordinated Federal approach to fellowships through cooperative effort of Federal agencies.	Efficient and effective Federal approach to providing fellowships for developing highly skilled S&E workforce, and preparing students in areas of national need.	Evaluation results that determine impact of Federal investments in fellowships and inform future investments.

Provide targeted opportunities for fellows' preparation in areas critical to the Nation and, in particular, critical to preparing the workforce needed to advance the missions of Federal agencies, including exploring scholarship-for-service and STEM-focused pathways to federal service programs throughout the Federal research and engineering enterprise.		
Actions	Outcomes	Metric or Milestone
Near-term (years 1-2)		
Develop mechanisms to increase the Fellows receiving opportunities for preparation in areas of national need, including, where appropriate, establishing interagency MOUs or other mechanisms of coordination between NSF and the other CoSTEM agencies .	Established process for identifying areas of national need and developing opportunities in those areas. Evidence of quality preparation	Evidence of quality preparation in areas of national need as assessed against goals of mission agency. Numbers of STEM fellowship program graduates serving in federal workforce. Progress on meeting national STEM workforce needs.
Mid-term (years 3-4)		
As appropriate, explore the enhancement of fellowship programs to create expedited pathways into public service for STEM professionals. Continued identification of opportunities in areas of national need.	Established infrastructure to offer Fellows development of opportunities in areas of national need	Increased number of opportunities in areas of national need offered to increased number of fellows.
Long-term:		
Management of Federal comprehensive approach to fellowships through cooperative effort of Federal agencies.	Highly efficient and effective Federal approach to providing fellowships for developing highly skilled STEM workforce, and preparing students in areas of national need.	Evaluation results that describe impacts of Federal investments in fellowships and inform future investments.

Continue and enhance mechanisms that evaluate the impact of fellowships to inform future Federal investments.		
Actions	Outcomes	Metric or Milestone
Near-term (years 1-2)		
Refine evaluation mechanisms to inform the improvement Federal fellowships for increased impact.	Where appropriate, longitudinal tracking system designed to provide useful information about impact of graduate experiences on preparation of STEM professionals.	Design of tracking system is accepted across Federal agencies.
Mid-term (years 3-4)		
Continued refinement and implementation of evaluation mechanism.	Implementation of evaluation mechanism.	Initial evaluation results on impact of fellowships toward developing highly skilled U.S. workforce, as well as workforce in areas of national need.

Actions	Outcomes	Metric or Milestone
Long-term:		
Coordinated Federal approach to fellowships through cooperative effort of Federal agencies.	Highly efficient and effective Federal approach to providing fellowships for developing highly skilled S&E workforce, and preparing students in areas of national need.	Evaluation results that determine impact of Federal investments in fellowships and inform future investments.

5.2. Implementation of the Coordination Objectives

Coordination Objective 1: Build New Models for Leveraging Assets and Expertise

Implement a Concept of Lead and Collaborating Agencies to Leverage Capabilities Across Agencies to Ensure the Most Significant Impact of Federal STEM Education Investments

Funding for Federal STEM education efforts has been dispersed across multiple agencies in many different programs and initiatives – a model that makes use of specific agency strengths and assets. With this Strategic Plan, a new model for this portfolio of investments is proposed, with a concept of lead and collaborating agencies organized around each of the priority goals.

Background

With more than 200 STEM investments among 14 agencies there are certainly programs and investments that have complementary elements that can be coordinated, overlapping elements that can be considered for consolidation, and potential synergies that can be maximized for stronger impact. The new approaches outlined in this plan will enhance the potential for scale, the impact, and the quality of the Federal STEM education investment in each of the strategic priority areas.

Lead agencies will be responsible for convening CoSTEM agencies and setting a coherent and detailed agenda for their area of responsibility, assessing the collaborative assets of other agencies as part of the plan, moving the implementation plans forward in the respective goal areas, and ensuring that planning for evaluation and impact studies occurs. These activities will be coordinated with the collaborating agencies that bring unique assets, knowledge, and facilities to the table, and where appropriate, assisted by the CoSTEM Implementation Subcommittees. More efficient leveraging of funding and other resources across agencies is a critical component of building coherence in the investment portfolio. Alignment of investments will lead to increased efficiency within and across agencies as well as clarification of capabilities, roles, and missions.

Significant Related Efforts

There are a number of venues in which Federal agency staff with responsibility for STEM education investments regularly come together to collaborate and address major national issues. These activities will serve as the baseline foundation for the new model. For example, the Informal Science Education Interagency Working Group was recently created to develop increased coordination and collaboration among Federal agencies. Other groups also exist, such as the NSTC National Initiative for Cybersecurity Education (NICE), the Networking and Information Technology Research and Development (NITRD) program, the Environmental Education Taskforce, the Ocean Education Interagency working group, the U.S. Global Change Research Program Communication and Education Working Group, and the OMB evaluation group.

In addition, because agencies tend to interact with overlapping stakeholder communities, the annual meetings held by such professional societies as the American Mathematical Society, the Society for Research on Educational Effectiveness, the National Association for Research in Science Teaching, the American Education Research Association, the Council of Graduate Schools, the American Association for the Advancement of Science, National Science Teachers Association, American Geophysical Union, Association of Science and Technology Centers, the Council of State Science Supervisors, and others often include sessions that bring together experts from several agencies to discuss program offerings and connections. Some key Federal committees that are concerned with STEM education matters, such as the Committee on Equal Opportunities in Science and Engineering (CEOSE) and the National Board for Education Sciences have designated agency representatives or liaisons who participate in their meetings. Finally, there are a number of collaborative undertakings in STEM education that involve two or three agencies working together (e.g., the NSF-ED K-16 Mathematics Initiative and the ED-NASA Summer of Innovation and 21st Century Community Learning Centers work). These efforts, together with the efforts of CoSTEM, all provide a strong context for reaching the next levels of collaboration.

Strategies

To build and implement new models for leveraging and collaborating across agencies, Federal agencies will explore key strategies including:

- Collaborating in a new arrangement of lead and collaborating agencies to build implementation roadmaps in the goal areas;
- Designing new infrastructure, networks, and mechanisms to ensure that assets and resources available through Federal agencies are used and accessed widely for the improvement of STEM learning across the country, and reducing administrative barriers to collaboration; and
- Developing a framework to guide coordinated CoSTEM agency budget requests.

Table of potential actions/outcomes/metrics

Collaborate in a new arrangement of lead and collaborating agencies to build implementation roadmaps in the goal areas.		
Actions	Outcomes	Metric or Milestone
Near-term (years 1-2)		
As appropriate lead and collaborating agencies will convene around relevant priority areas. Where appropriate, CoSTEM will create new Implementation subcommittees in the strategic priority areas.	Agreements among lead and collaborating agencies about detailed agenda and plans for progress in each strategic priority area.	Roadmap and detailed implementation timelines.

Actions	Outcomes	Metric or Milestone
Mid-term (years 3-4)		
Track progress on implementation, collect evidence, and recommend mid-course corrections.	Evidence-based planning for appropriate adjustment and updating of strategic priorities and objectives in each area.	Implementation subcommittees could provide input to inform development of subsequent Administration STEM education budget requests.
Long-term:		
Recommend additional priority areas for longer-term improvement in STEM education. Take steps to assess effectiveness of this approach.	Continuous assessment of needed improvements in STEM education based on evidence in implementation in original five priority areas.	Concept papers developed or commissioned to assess needed emphases and priorities for long-term improvement of STEM education.

Design new infrastructure, networks, and mechanisms to ensure that assets and resources available through Federal agencies are used and accessed widely for the improvement of STEM learning across the country, and reduce administrative barriers to collaboration.		
Actions	Outcomes	Metric or Milestone
Near-term (years 1-2)		
Assemble information about government mechanisms that relate to cross agency collaboration and recommend approaches for streamlining and simplifying to promote collaboration. Benchmark existing federal infrastructure, networks, etc.	More capacity to collaborate across agencies based on shared understanding of procedures and mechanisms.	Simplification of key processes such as development of MOUs and administrative clearances to encourage common procedures and collaborations.
Mid-term (years 3-4)/Long-term:		
Introduce innovative strategies for cross-agency capacity building and expertise development in STEM education.	Such innovations as shared personnel, common offices that span agencies, and joint funding opportunities.	An increased in joint funding opportunities that include more than two agency partners.

Develop a framework to guide coordinated CoSTEM agency budget requests.		
Actions	Outcomes	Metric or Milestone
Near-term (years 1-2)		
As appropriate, convene CoSTEM agency representatives, including CFOs, to determine feasibility, timing and process for collaborative budget requests, potentially as early as the FY 2015 request.	Coordinating budget process for STEM education.	Production of cross-agency STEM education budget request framework.
Mid-term (years 3-4)/Long-term:		
Assess impact and effectiveness of the coordinated request framework and revise practices and approach accordingly.	Revised coordinated budget process for STEM education.	Continued collaboration on cross-agency coordinated STEM education budget request.

Coordination Objective 2: Build and Use Evidence Based Approaches

Conduct Rigorous STEM Education Research and Evaluation to Build Evidence About Promising Practices and Program Effectiveness, Use Across Agencies, and Share with the Public to Improve the Impact of the Federal STEM Education Investment

An ever-growing body of education research and evaluation findings is improving our understanding of STEM learning and STEM instructional practices. These include a wide set of studies, including foundational work on the brain and cognition research on the critical role that "non-cognitive" factors play in aiding achievement in STEM fields, examinations of classroom design, implementation research on teacher education programs, and evaluations of large-scale education reform efforts. Findings from emerging research are more useful when they are synthesized and shared and when the studies build from existing evidence. In addition, gaps between fundamental research and studies to understand and validate promising practices persist.

Background

Federal STEM education programs are more likely to have a greater impact if they use, amass, and improve evidence of what works. There are substantial bodies of education research and evaluation studies that have influenced the general directions of improvement in STEM education. Basic findings about how people learn STEM and how to teach STEM subjects,[113, 114] for instance, are consistent with current emphases on active engagement with meaningful STEM learning settings. Similarly, there is a large body of work in the area of STEM education evaluation, at the level of projects, lines of activity (e.g., IES' evaluation of teacher professional development[115]), and full Federal programs/investments. Some evaluation methodologies also lead to insights about why, for whom, and under what conditions particular interventions or programs are effective or not. At the same time, there is substantial area expertise and experience within the STEM education staffs of the CoSTEM agencies that is deployed with the planning, revision, and implementation of any Federal investment. This professional knowledge has been refined over time and can serve to guide the effective planning and implementation of investment activities, including the design principles described in Appendix B.

Despite a rich body of work, more can be done to leverage the findings of research studies and evaluation efforts to strategically shape the direction of Federal investments. Given limited resources and the availability of some evidence of the most promising STEM education practices, it is important that those shaping government investments incorporate existing knowledge to make investments most likely to have the intended impact. At the same time, it is important to decide which research and evaluation questions can be addressed with new investment ideas, so the body of evidence continues to grow and our collective understanding is both improved and communicated.

Collaborative strategies for evaluating individual investments within and among agencies and themes that cut across several investments can improve the evaluation capacity of all agencies. Models such as the MIT Poverty Action Lab, or creative use of administrative data to undertake impact studies, as NIH and others have done [116, 117] should be explored. In addition, agencies might consider developing cross-agency monitoring systems that will allow the collection, comparison, and analysis of uniform data across investments.

Significant Related Efforts

Syntheses of research are produced regularly by the National Research Council; recent examples, supported by NSF, have focused on K-8 science learning education,[118] learning science in informal environments,[119] STEM schools,[120] monitoring progress in STEM education,[121] discipline based education research,[122] and improving minority participation in STEM.[123] The most recent major synthesis of research on mathematics learning K-8 is the 2008 report of the President's National Math Advisory Panel.[124] ED has supported syntheses on many topics, including research on teacher preparation.[125] The Institute of Education Sciences (IES) has a searchable database that includes several funded synthesis studies; in addition, the IES-sponsored What Works Clearinghouse contains reviews of many interventions. Recent PCAST reports [126, 127] provide recommendations that are based on such research findings. Synthesis of evidence-based practices and research findings is an ongoing process, and additional syntheses are needed periodically as empirical findings continue to emerge. In addition, the National Research Council has reviewed agency education portfolios, providing valuable perspective on these efforts.[128, 129]

Agencies have also made strides in developing their evaluation capacity. For example, ED's IES has made rigorous, experimental, and quasi-experimental evaluation the norm in many education issue areas. Further, ED's Investing in Innovation program spearheaded a "tiered evidence" structure that tests new ideas, validates promising interventions, and scales up proven practices. Additionally, NSF has created a STEM Education Evaluation Working Group to improve evaluation of its own STEM education programs. Other agencies have been making continual improvements to their evaluation capacity during recent years and have completed agency-wide plans for evaluation of STEM education investments. As another example, DOI completed its first STEM education and employment strategic plan, which provides guidance that encourages a culture of evaluation to foster continual improvements and impact assessments. NOAA has embarked on an ambitious monitoring and evaluation effort to assess its education outcomes and impacts across programs. The purpose of this effort is to link outcome measures to program efforts and to evaluate the success of the agency in meeting its strategic goals.

Research and evaluation efforts also yield important methodologies and instruments that can be shared across government for use within research and evaluation efforts, such as the *Framework for Evaluating Impacts of Informal Science Education Projects*[130] produced through a multi-agency collaboration. The development of other instruments and tools is supported by programs including IES' Statistical and Research Methodology in Education grant program, and the NSF Promoting Research and Innovation in Methodologies for Evaluation (PRIME).

Agencies have already begun to develop shared strategies for generating evidence. ED and NSF have been collaborating on the development of a common evidence framework. The guidelines include suggestions for describing high-quality work within six education research study types: (1) foundational research; (2) early stage and exploratory research; (3) design and development projects; (4) efficacy studies; (5) effectiveness studies; and (6) scale-up studies.[131] The evidence guidelines are intended to be used by prospective grantees, Federal agency staff members, and merit reviewers.

The design principles developed by CoSTEM (see Appendix B) represent a synthesis of knowledge from research and practice about necessary conditions for effectiveness in each investment area. CoSTEM posted the design principles for public comment[132] and made adjustments based on the feedback received. Applying these principles to program design and implementation is intended to improve the prospects of accomplishing an investment's objectives and enhance cross-agency coordination.

Over the past several years, the Administration has laid a strong foundation and developed new tools that will help Federal agencies support evidence-based practices or strategies that will build useful knowledge of what works. For example, the Administration has encouraged the use of evidence-based grant program designs,[133] and identified approaches that can inform the development and continuous refinement of Federal grant-making. The guidance is based on the theory that program design and implementation strategies should build evidence about what works to increase Federal resources flowing to state, local, or other grantees using evidence-based practices and to identify more cost-effective practices. For example, a growing number of STEM education investments are implementing strategies such as tiered evidence frameworks.

Strategic Objectives

To improve the exploration and sharing of evidence-based practices the Federal agencies will undertake the following strategic objectives:

- Support syntheses of existing research that can inform Federal investments in the STEM education priority areas;
- Improve and align evaluation and research strategies and expertise across Federal agencies, with lead agencies helping lead design of evaluation in their assigned goal areas; and
- Lower barriers to interagency collaboration by streamlining processes for interagency collaboration (e.g., MOUs, Interagency Agreements, shared staff, common instruments, etc.).

Table of potential actions/outcomes/metrics

Support syntheses of existing research that addresses critical issues in STEM education priority areas.		
Actions	Outcomes	Metric or Milestone
Near-term (years 1-2)		
Identify critical research questions in each priority area working with stakeholder communities, and commission syntheses as appropriate.	Research synthesis and agenda in each priority area.	A series of research syntheses and agendas developed in collaboration with stakeholders.
Mid-term (years 3-4)		
As appropriate, agency investments are leveraged to address critical research questions.	Release requests for proposals to address critical research questions. NSF and ED will align some of their calls for research applications and proposals with critical research questions.	Funded research and evaluation studies and activities in the priority areas are assembled as a cross-agency portfolio and a monitoring system is introduced.

Long-term:		
Support symposia for each priority area with participation from stakeholders, researchers, practitioners, and Federal agency representatives.	Convening and consolidation of findings and sharing of ideas across multiple communities.	Analysis of literature and funded efforts in priority areas to examine nature of the knowledge base.

Improve and align evaluation and research strategies and expertise across Federal agencies.		
Actions	Outcomes	Metric or Milestone
Near-term (years 1-2)		
As appropriate, create a cross-agency STEM Education Evaluation working group to assess capacity and practices.	Agreement to inter-agency support of evaluation capacity in STEM education.	As appropriate, mechanisms for sharing STEM education evaluation capacity across agencies are established.
Mid-term (years 3-4)		
Agencies work together to implement a major STEM education investment evaluation for an investment in one of the priority areas that can provide evidence of impact, using current best practices and methodologies in evaluation.	A coordinated set of STEM education impact evaluations in the priority areas.	Agencies will collaborate in the implementation of impact evaluations in the four priority areas.
Long-term:		
Agencies develop monitoring systems to continue to document progress in priority areas.	New monitoring systems and new data accumulation.	Coordinated data systems and open access to information. Link to indicator and monitoring systems.

Lower barriers to interagency collaboration by streamlining processes for interagency collaboration (e.g., MOUs, Interagency Agreements, shared staff, common instruments, etc.).		
Actions	Outcomes	Metric or Milestone
Near-term (years 1-2)		
Agencies adapt Design Principles to fit context and shares across CoSTEM agencies.	Internal agency discussions about the process and considerations in investment design.	Revised Design Principles, incorporating agency details, made available across agencies.
Mid-term (years 3-4)		
Agencies introduce processes for documenting the use of Design Principles and evidence-based practices in the development of new STEM education solicitations or requests for proposals.	Internal change of culture in agency discussions about use of Design Principles and evidence in program planning.	New solicitations or requests for proposals incorporate in visible ways considerations of design principles and evidence.
Long-term:		
Cross-agency evaluation is planned to examine impacts of using design principles and evidence in development and revision of programs.	Documentation of how Design Principles and Evidence were used.	Process for tracking the impact of using Design Principles and evidence is in place.

5.3. Implementation Constraints

It is worth acknowledging external factors beyond the direct control of Federal agencies that may impede achievement of the goals in the Strategic Plan. Annual appropriations decisions by Congress will directly impact the flexibility that agencies have to coordinate and evaluate their efforts.

CoSTEM agencies have noted a number of potential constraints to successful implementation of this plan, including, among others:

- Overall agency budget fluctuations make long-term planning difficult.
- In some cases, authorizing language of some departments and agencies and sub-agencies limits them from targeting underrepresented groups.
- Challenges associated with collecting and sharing student data can limit STEM education program-evaluation strategies.
- For the most part, infrastructure and expertise for STEM education investments varies widely across departments and agencies.

CONCLUSION

Given that many jobs of the future will be STEM jobs, that our K-12 system is "middle of the pack" in international comparisons, and that progress on STEM education at multiple levels is critical to building a just and inclusive society, there is an urgent need to continue to improve STEM education in the United States. Much knowledge for how to make progress exists. Federal agencies have the collective capacity to make a difference. This Strategic Plan identifies goals, priorities, and a new framework and mechanisms for collaboration and program improvement. This Strategic Plan charts a course for sustained improvement and can help Federal investments in STEM education programs make a difference for many more students, educators, and members of the public. Implementation will require making hard choices, forging new partnerships, and focusing on outcomes. Adjustments along the way will be necessary but the Federal agencies will work together to make progress on this national priority.

APPENDIX A

The Federal Science, Technology, Engineering, and Mathematics (STEM) Education Portfolio: FY 2011 [134]

This Appendix details the results of CoSTEM inventory of Federal STEM education investments in FY 2011. The Fast-Track Action Committee on Federal Investment in STEM Education (FI-STEM) was chartered in April 2010 to develop the inventory process, analyze the inventory results, and draft a portfolio report and this update with the oversight of CoSTEM. The membership of the committee included representatives from the 12 Federal agencies that comprise CoSTEM. This appendix updates the original (FY 2010) inventory published in December 2011 as *The Federal Science, Technology, Engineering, and Mathematics (STEM) Education Portfolio* with an FY 2011 inventory. This inventory includes detailed information on STEM education investments in order to identify potential areas of synergy across and within agencies; support sharing of effective STEM education investment strategies and evaluation techniques across Federal agencies; increase awareness of STEM education investments within and across Federal agencies; and support development of a Federal STEM education 5-year strategic plan that is the main body of this report. The full FY 2011 inventory is Appendix Table A4.

The inventory process is also a catalyst for interagency collaborations related to improving investments in STEM education. By enhancing capabilities of the interactive database housing the inventory, CoSTEM anticipates that the inventory will serve as a tool to bring together staff working on similar efforts, facilitate their collaboration, and support their ability to implement program improvements.

Executive Summary of Appendix A

Over the past year, the NSTC's Committee on Science, Technology, Engineering, and Mathematics (STEM) Education (CoSTEM) completed an inventory of FY 2011 Federal investments in STEM education. This appendix is a report of the findings of that inventory which, together with the original FY 2010 inventory, informed development of the Strategic Plan that is the body of this report.

Representatives from the 12 CoSTEM agencies developed the online survey used to collect the information. The FY 2011 survey was similar to the one used for the FY 2010 portfolio report.[135] Changes made to the FY 2010 inventory survey improved its clarity and the accuracy of the information collected.

The inventory data indicate that in FY 2011 the Federal Government expended $2.891 billion on more than 200 STEM-focused investments. Almost 80 percent of FY 2011 Federal STEM education funding was distributed through three Federal entities: the National Science Foundation (NSF) (40 percent); the Department of Health and Human Services (HHS) (20 percent); and ED (19 percent).

Additional findings from the analysis of the FY 2011 Federal inventory were useful for the development of the Strategic Plan, including the following:

- About one-third of Federal STEM education funding was spent on activities that target the specific workforce needs of Federal agencies surveyed, and the remaining two-thirds targeted broader STEM education activities.
- More than half of the funding (52 percent) was devoted to investments with a primary objective related to postsecondary STEM education (37 percent with a "STEM Degrees" objective; 15 percent with a "STEM Careers" objective).
- Thirty percent of funding went to investments that focused *primarily* on supporting underrepresented groups and that placed a competitive priority on supporting these groups, while an additional 60 percent of funding supported investments that *encourage* serving groups historically underrepresented in STEM fields, but do not make it a competitive priority.

- About 10 percent of Federal STEM education funding was dedicated to investments with the primary goal of supporting the preparation, education, or professional development of STEM teachers. An additional 33 percent of funding supported STEM teachers as a secondary objective.

The FY 2011 and FY 2010 portfolio reports provide a baseline CoSTEM can use to analyze funding trends and to identify areas potentially needing more investment.

The Federal STEM Portfolio: FY 2011

The *Federal Science, Technology, Engineering, and Mathematics (STEM) Education Portfolio: FY 2010* was the first of two interagency efforts to systematically survey and analyze data about Federal activities in science, technology, engineering, and mathematics (STEM) education. The inventory gathered information about how much the Federal Government spends on STEM education investments overall, and how many and what kinds of investments individual Federal agencies support. This examination of the FY 2011 Federal STEM education portfolio notes changes in investments from the previous year and examines the survey's impact on investment coordination and improvement.

CoSTEM chartered the NSTC Fast-Track Action Committee on Federal Investment in STEM Education (FI-STEM) in April 2010 to develop and administer the inventory. CoSTEM oversees the activities of FISTEM, which includes members from the 12 Federal agencies represented on CoSTEM. FI-STEM met eight times between December 2011 and September 2012 to discuss outcomes of the previous inventory, revise criteria for the current inventory, update survey structure and definitions, determine collection and analysis methods, and develop a process for drafting this inventory report.

Criteria for Inclusion

For the inventory, detailed information was collected on Federally-funded STEM education investments in FY 2011 that had current and consistent funding at or above \$300,000 annually (see Table A1 for the full criteria set). Only basic information such as budget, agency, investment name, evaluations conducted, focus on historically underrepresented groups, and funding mechanism was collected on STEM programs with budgets of less than \$300,000, and/or that were funded by earmark or with American Recovery and Reinvestment Act funds. The decision to collect detailed information on only investments at or above \$300,000 level was based on the judgment of FI-STEM members.

Table A1 details definitions, units of analysis, and other inclusion criteria used in the inventory. Other types of agency activities may contribute to STEM education and have been included in prior surveys (such as the *Report of the American Competitiveness Council*[136] and *Survey of DOD STEM Programs*[137]) but fall outside the scope of CoSTEM inventory. They include:

- research investments that fund science or engineering research and can support undergraduate or graduate students who assist in carrying out research if scientific research produced is the primary goal and measure of success;
- general education investments (such as Pell or Title I grants) that may include STEM as one of several education topics;
- volunteer activities by agency staff, such as classroom visits or judging STEM competitions, that do not require Federal funds;
- investments to promote public awareness of STEM education investments; and
- postdoctoral research awards or fellowships that support scientific research.

STEM education investments by the 12 CoSTEM agencies and the Nuclear Regulatory Commission (NRC) met criteria for inclusion in the inventory. The most common reasons that the activities were not included in the inventory were that the budget for the activity was under $300,000 or that the activities were part of a larger education investment that supported education in many fields other than STEM.

Table A1. Criteria and Definitions

STEM: For the purposes of this inventory, STEM includes physical and natural sciences, technology, engineering, and mathematics disciplines, topics, or issues (including environmental science education or environmental stewardship). We recognize that various different and usually broader definitions are used for "STEM." This relatively narrow definition has been chosen to constrain the focus of the inventory to specific areas that have similar educational contexts, issues, and challenges, in order to maximize the inventory's usefulness in characterizing and improving the effectiveness of the Federal spending intended to address this particular set of educational contexts, issues, and challenges.

Investment (the unit of analysis in the detailed survey): A funded STEM education activity that has a dedicated budget of or above $300,000 (potentially part of a budget for a larger program, but excluding a onetime or irregular expenditure of overhead funds and salary and expenses costs), staff to manage the budget, and was funded in FY 2011.

STEM Education: Formal or informal, in-school or out-of-school, education that is primarily focused on physical and natural sciences, technology, engineering, and mathematics disciplines, topics, or issues (including environmental science education or environmental stewardship). All the investments included in this STEM education inventory have one of the following as a primary objective:

- **Learning:** Develop STEM skills, practices, or knowledge of students or the public.
- **Engagement:** Increase learners' interest in STEM, their perception of its value to their lives, or their ability to participate in STEM.
- **Pre- and In-Service Educator or Education Leader Performance:** Train or retain STEM educators (K– 12 pre-service or in-service, postsecondary, and informal) and education leaders to improve their content knowledge and pedagogical skills.
- **Postsecondary STEM Degrees:** Increase the number of students who enroll in STEM majors, complete STEM credentials or degree programs, or are prepared to enter STEM careers or advanced education.

- **STEM Careers**: Prepare people to enter into the STEM workforce with training or certification (where STEM-discipline-specific knowledge and skill are the primary focus of the education investment).
- **STEM System Reform**: Improve STEM education through a focus on education system reform.
- **Institutional Capacity**: Support advancement and development of STEM personnel, programs, and infrastructure in educational institutions such as universities, informal education institutions, and state and local education agencies.
- **Education Research and Development:** Develop evidence-based STEM education models and practices.

Survey Structure

The inventory classifies Federal agency STEM education investments into two broad categories:

Agency-mission workforce education investments develop or train the STEM workforce of the agency or the workforce in fields directly related to the agency's mission (such as in aerospace engineering, national security science, and nuclear regulatory science). These typically include graduate scholarships, undergraduate internships, or institutional capacity-building in fields or degrees tightly aligned to an agency's mission.

Broader STEM education investments support formal and informal STEM education not focused on a specific discipline, STEM education research, and STEM education capacity building to improve interest in and understanding of STEM concepts and to enhance the broader national STEM workforce.

The FY 2011 survey contains five sections:

- Section 1: Background information (name, agency, and primary staff contact)
- Section 2: Descriptive information about broader STEM education investments (objectives,
- audiences, and numbers served)
- Section 3: Funding information about broader STEM education investments[138]
- Section 4: Evaluation information about broader STEM education investments
- Section 5: Information about agency-mission workforce education investments

Survey Data and Analysis

The FY 2011 inventory identified more than $2,891 million across more than 200 STEM education investments from 13 Federal agencies (the 12 CoSTEM member agencies and the NRC) (Figure A1). The investments include a range of programs and activities, such as student scholarships, grants to institutions, and contracts with STEM education organizations. Some investments also generate products and services for students and educators.

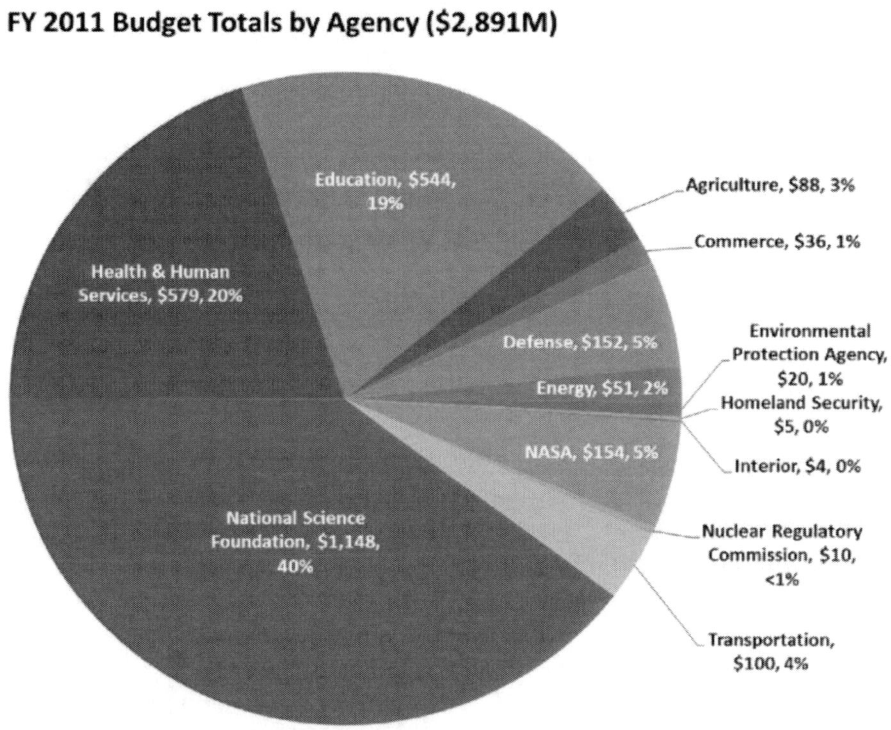

Figure A1. Federal Stem Education Investments >$300k, By Agency (In Millions).[139]

Investment Focus

Two-Thirds of Investments Focus on Broader STEM Education

In FY 2011, Federal agencies spent $1,903 million on education investments focused on broader STEM education (66 percent of the nearly $2,891 million total) and $988 million (34 percent) on agency-mission workforce education.

The Federal STEM education portfolio reflects the agencies' missions and evolution of their roles in STEM education, which is influenced by the history and goals of education funding at each agency, emerging national needs, agency missions, Congressional and Presidential direction, and each agency's unique assets and capabilities. ED and NSF share the broadest missions in education and STEM education research respectively. ED and NSF together sponsor the majority of Federal investments in broader STEM education. Nearly all their investments focus on broader STEM issues and activities.

Most Agencies Focus the Majority of Their Investments on Agency-Mission Workforce Issues

Science-mission agencies have broad responsibilities and make strategic investments in educating their workforces. Science-mission agencies devote between 55 and 100 percent of their STEM education investments to agency-mission workforce education.

Some agencies, including USDA, DOC, DOD, NOAA, and NASA, support both broader STEM education and agency-mission workforce education investments. Other agencies, such as DOE, HHS, and DOT, invest almost exclusively in mission-workforce development and training, whereas, DHS, DOI, and NRC allocate investments entirely to the education of a mission-related workforce.

K–12 Education Receives Largest Share of Broader *STEM* Investments

About $1.1 billion in broader STEM education investments were directed toward improving STEM education for K–12 students or teachers in FY 2011. About 60 percent of investments in broader STEM education focus on two or more types of education activities (e.g., K–12 education and undergraduate education; or informal education, K–12 education, and graduate education).

Primary Objectives

More Than One-Third of Investments Are Aimed at Helping Students Obtain Postsecondary STEM Degrees

Agencies identified a single primary objective (from a list of eight) for each investment (see Table A1). "Primary objective" was defined as the main desired outcome or basis for evaluating that investment under ideal circumstances. The Postsecondary STEM Degree primary objective accounted for 37 percent of all funding, followed by Education Research and Development (R&D; 16 percent) and STEM Careers (15 percent). Investments in STEM Careers and Postsecondary STEM Degrees are more likely to be related to agency-mission workforce education than to broader STEM education.

Funding for investments with the primary objective of Engagement was $164 million. In addition, engagement is the secondary objective of many investments. In fact, $1,444 million was spent on investments where Engagement was a secondary objective.[140]

About 70 percent of investments with the Postsecondary STEM Degrees primary objective were accounted for by five broader STEM education programs supported by NSF. They are the Graduate Research Fellowship Program; Scholarships in Science Technology, Engineering and Mathematics; Advanced Technological Education; Integrative Graduate Education and Research Traineeship Program; and Louis Stokes Alliances for Minority Participation. They included a range of investment types, from providing direct support for students seeking postsecondary STEM degrees, to providing support to 2- year and 4-year colleges, to supporting innovative models of graduate education and STEM-degree attainment among individuals from underrepresented groups. HHS investments accounted for nearly 40 percent of the funding for agency-mission workforce investments with the primary objective of Postsecondary STEM Degrees. HHS investments provided direct support to students or to 2-year and 4- year colleges to ensure sufficient graduates in biomedical fields.

$1,265 Million in Investments Have an Objective Aimed at Educator Performance

Investments with the primary objective of Pre- and In-Service Educator and Education Leader Performance (Educator Performance) accounted for $315 million. Seventy-eight percent of those funds supported professional development (PD) for in-service educators, while the remainder supported both teacher PD and pre-service educators. No Federal investments are dedicated solely to funding efforts for pre-service educators. Educator

Performance is listed as a secondary objective in investments that account for about $950 million.

Serving Groups that are Underrepresented and Underserved in STEM

Promoting equitable participation of people seeking STEM degrees and careers in STEM is a Federal priority and a common emphasis of investments in this survey. Information was collected on groups historically underrepresented in STEM fields, including Hispanics and Latinos, African Americans, American Indians, Alaska Natives, Native Hawaiians, and Pacific Islanders. In addition, information was collected on groups that may be disadvantaged or underserved in specific fields of STEM such as women, people with disabilities, economically disadvantaged people, and those living in rural or urban areas.

The FY 2011 survey used several questions to quantify investments that seek to enhance participation of people from underrepresented and underserved groups in STEM education. However, interpreting these data is difficult because agencies follow different regulations for allocating funds for programs directed toward specific groups. Further, statutory authority sometimes determines how, or whether, Federal agencies can legally limit specific investments to certain populations, such as through competitive priority or by setting aside specified funding amounts for specific populations.

Postsecondary Degrees and Institutional Capacity are top objectives for STEM equity

Eighty-nine percent of investment dollars were reported to emphasize or encourage support of underrepresented groups. Table A2 shows primary objectives of investments and type of focus. Of the $616 million in investments that focused primarily on underrepresented groups, 40 percent focused primarily on Postsecondary STEM Degrees ($248.5 million) and 33 percent focused on Institutional Capacity ($201.5 million). Postsecondary STEM Degrees was the most-funded of the eight primary objectives ($1,065 million), accounting for 37 percent of those investments.

Table A2. Distribution by Primary Objective of Investments Targeting Groups Underrepresented and Underserved in STEM, by Type of Focus on Underrepresented Groups (in millions)

Type of Focus on Underrepresented Groups	Education R&D	Engage-ment	Institutional Capacity	Learning	STEM Degrees	Pre- and In-Service	STEM Careers	Total
Primary Focus	$27	$48	$201	$22	$249	$0	$68	$616
Specific Percent of Funding	$0	$8	$0	$0	$0	$0	$0	$8
Competitive Priority	$195	$7	$2	$16	$75	$1	$1	$296
Highly Encouraged	$228	$88	$9	$135	$610	$246	$328	$1,644
No Focus	$7	$13	$5	$71	$130	$69	$32	$327
Total	$457	$164	$217	$245	$1,065	$315	$429	$2,891

Institutional Capacity was a less common primary focus overall, yet 33 percent of investment dollars that focus primarily on underrepresented or underserved groups support building Institutional Capacity.

Minority Serving Institution Investments account for 10 percent of Federal STEM education funding

More than 20 Federal STEM education investments targeted historically unrepresented groups by supporting minority serving institutions (MSI). Overall, $297 million was directed towards MSIs. Twenty-eight percent can be awarded to any type of MSI. The remainder was targeted to specific types of MSIs, such as Historically Black Colleges and Universities (HBCU), Hispanic-Serving Institutions (HSI), Alaska Native-Serving Institutions, Native Hawaiian-Serving Institutions, and Tribal Colleges and Universities. The largest amount of funding ($110 million) was directed to HSIs, mostly from ED's Developing Hispanic-Serving Institutions STEM and Articulation Program ($100 million).

Audiences Served

Pre-K Through Undergraduate Learners Are the Largest Audience Served

Federal STEM education investments served an array of audiences, and some served more than one. As illustrated in Table A3, the largest amount of funding ($2,354 million) is for investments that served learners in pre-kindergarten through graduate school. Investments that served educators and staff at formal education institutions are secondary in both number and amount of investments. Education researchers and informal-education educators and researchers received fewer funds; programs that served adults receive the least ($325 million).

Table A3. Education Investments, by Target Group (In Millions)[141]

Audience	FY 2011 Total
Learners Aged Pre-K to Grade 20	$2,354
K–12 Classroom Teachers	$1,059
K–12 Staff, Leaders, and Administrators	$416
Postsecondary Instructors	$837
Postsecondary Dean, Leaders, and Administrators	$371
Education Researchers	$386
Informal STEM Educators	$367
Informal STEM Education Leaders and Program Developers	$341
Adults[142]	$325

Interagency and Non-Federal Collaboration[143]

Thirty-two percent ($610 million) of the funding for broader STEM education supported collaborations with other Federal agencies or non-Federal groups. About $500 million supported interagency collaborations, including $123 million that supported collaboration across agencies and with non-federal organizations. Collaboration mechanisms include memoranda of understanding, interagency funding solicitations, and shared expertise.

Evaluation

Questions in both the FY 2010 and FY 2011 survey examined evaluations at the investment level. In FY 2011, 104 investments required grantees or investment components to conduct evaluations.

Of the more than 200 investments, 164 have been evaluated; 144 of those indicated the year the most recent evaluation was conducted. Since 2005, at least one evaluation has been completed for each of 138 investments (59 percent), with evaluations having been conducted on 44 percent in the past two years. A range of evaluation strategies was used at the investment level, and some investments were evaluated in more than one way.

Evaluation Designs Vary at the Investment and Component or Grantee Levels

For 109 investments their evaluation design was reported (Figure A2). Of those, the most common design was a pre-post intervention comparison with matched groups (46 investments, 29 percent). Randomized controlled trials were conducted for seven investments (4 percent). Fifty-five investments (34 percent) either did not report or did not specify an evaluation design. Forty-eight of the 109 investments used multiple evaluation designs, including four investments that used three different designs.

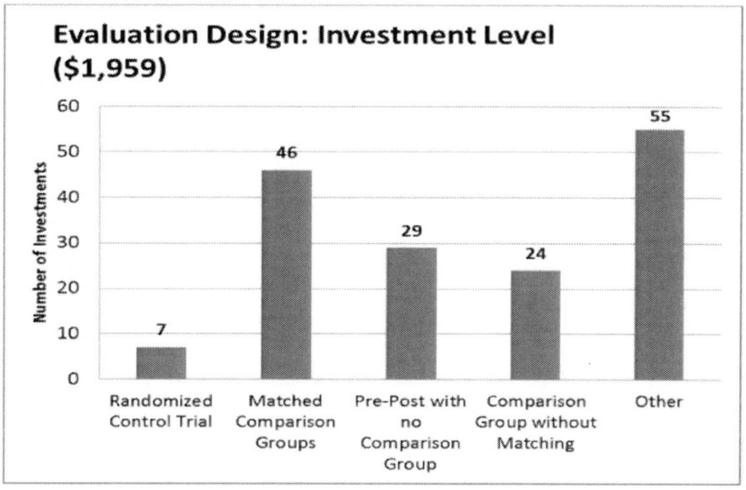

Figure A2. Number of Evaluations, by Evaluation Design (At the Investment Level).[144]

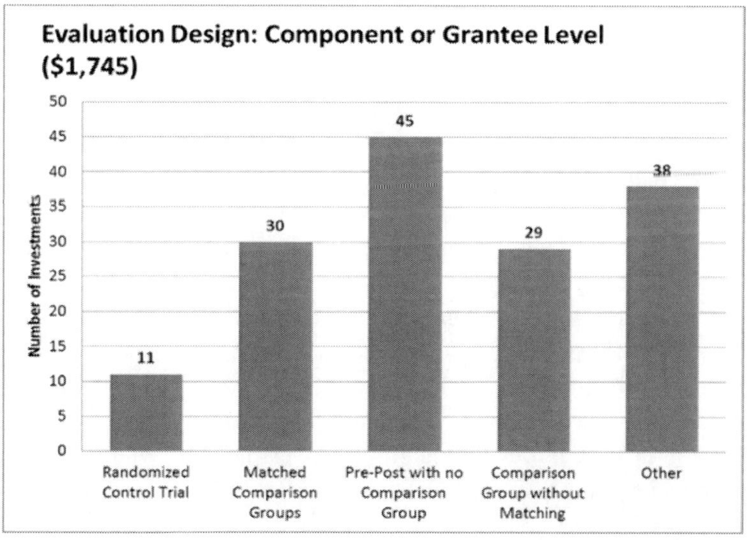

Figure A3. Number of Investments Since FY 2005, by Evaluation Design (At the Component Part or Grantee Level).

One-hundred-four investments required grantees or components to conduct evaluations. A variety of strategies were used at the investment component level, including those that identify how investments can be improved (formative) or test their impact (summative). Seventy-seven investments reported on the types of evaluation designs conducted at the level of the grantee or investment component (Figure A3). In contrast to evaluations at the investment level, the most common evaluation design at the grantee or component level was the pre-post assessment with no comparison group (45 investments). Few investments (11) reported evaluations using randomized designs at the grantee or investment or component level. Thirty-five of the 77 investments had components or grantees that used multiple evaluation designs, including 24 investments that reported using of 3 different evaluation designs.

Impact of the Federal STEM Inventory Process and the FY 2010 and FY 2011 Portfolio Reports

Benefits from Collecting a STEM Education Inventory

The STEM education inventories of FY 2010 and FY 2011 have enabled agencies to respond better to inquiries about their STEM education investments. In addition, agencies have used inventory results to drive development or revisions of their own internal STEM education strategic planning.

The inventory has also led to changes in agency coordination and collaboration by catalyzing additional partnerships between and among agencies.

Costem Can Help Promote Exchange of Federal Agency Staff Expertise

The FY 2011 survey asked each program director to identify areas of expertise (evidence-based instructional practices and scaling up of effective efforts, for example) to share with colleagues across the Federal Government, as well as areas in which they would likely benefit from other expertise.

Identifying and connecting to subject matter experts (115 investments) was the most-frequently reported need cited by Federal staff who completed the survey, followed by design efforts for underrepresented or underserved groups (94 investments) and peer-review best practices (92 investments). Less frequently reported areas of expertise included evaluation of informal education (26 investments), evidence-based instructional practices for informal education (29 investments), and promotional campaigns for education products (30 investments).

Agency staff (representing 108 investments) most frequently expressed a desire to learn more about "Design Efforts for Underrepresented or Underserved Groups." That was followed by expertise on summative evaluations (99 investments) and the use of social media and social networking (95 investments). The least frequently cited areas of needed expertise included request for proposals development (26 investments), conducting proposal merit-reviews (28 investments), and running STEM competitions (32 investments).

Appendix Table A4. List of Investments in the FY 2011 Inventory

Agency	Investment	FY 08	FY 09	FY 10	FY 11	Type	Primary Objective	Under-represented Groups
Agriculture	1890 Institutions Capacity Building Grants Program: Extension	4.53	5.00	6.08	6.40	Broader STEM	Institutional Capacity	Yes
Agriculture	Agriculture in the Classroom	0.55	0.55	0.55	0.60	Broader STEM	Pre- & In-Service Educator Performance	No
Agriculture	Alaska Native-Serving and Native Hawaiian-Serving Institutions Education Competitive Grants Program	3.20	3.20	3.20	3.20	Agency Mission Workforce	Institutional Capacity	Yes
Agriculture	Resident Instruction Grants Program for Institutions of Higher Education in Insular Areas	0.75	0.80	0.90	0.86	Agency Mission Workforce	Institutional Capacity	Yes
Agriculture	Distance Education Grants for Institutions of Higher Education in Insular Areas (DEG)	-	-	0.75	0.70	Agency Mission Workforce	Institutional Capacity	Yes
Agriculture	1890 Institutions Capacity Building Grants Program: Teaching	4.53	5.00	6.08	6.40	Agency Mission Workforce	Institutional Capacity	Yes
Agriculture	1890 Facilities Grant Program	17.27	18.00	19.77	19.77	Agency Mission Workforce	Institutional Capacity	Yes
Agriculture	Hispanic-serving Institutions Education Grants Program	6.05	6.24	9.24	9.22	Agency Mission Workforce	Institutional Capacity	Yes
Agriculture	AITC Secondary Postsecondary Agriculture Education Challenge Grants (SPECA)	0.98	0.98	0.98	0.98	Agency Mission Workforce	Learning	No
Agriculture	Higher Education Challenge Grants (HEC)	5.39	5.65	5.65	5.64	Agency Mission Workforce	Learning	No

Appendix Table A4. (Continued)

Agency	Investment	FY 08	FY 09	FY 10	FY 11	Type	Primary Objective	Under-represented Groups
Agriculture	New Era Rural Technology Competitive Grants Program (RTP)	-	0.75	0.88	0.87	Agency Mission Workforce	Postsecondary STEM Degrees	No
Agriculture	Higher Education Multicultural Scholars Program (MSP)	0.98	0.98	1.24	1.20	Agency Mission Workforce	Postsecondary STEM Degrees	Yes
Agriculture	Women and Minorities in Science, Technology, Engineering and Mathematics Fields Program (WAMS)	-	-	0.40	0.36	Agency Mission Workforce	STEM Careers	Yes
Agriculture	4-H Science, 4-H Youth Development Program	26.37	26.32	24.28	24.03	Broader STEM	Learning	No
Agriculture	National Institute of Food and Agriculture (NIFA) Fellowship Grants Program	6.00	6.00	6.50	3.00	Agency Mission Workforce	STEM Careers	No
Agriculture	Food and Agricultural Sciences National Needs Graduate and Postgraduate Fellowship Grant Program	3.68	3.86	3.86	3.90	Agency Mission Workforce	STEM Careers	Yes
Agriculture	AgDiscovery	0.49	0.49	0.49	0.50	Agency Mission Workforce	Engagement	No
Agriculture	1890 National Scholars Program	-	-	-	0.30	Agency Mission Workforce	Postsecondary STEM Degrees	Yes
Agriculture	**Total**	**80.75**	**83.82**	**90.85**	**87.94**			
Commerce	NOAA Bay Watershed Education and Training (B-WET)	9.55	9.70	9.70	2.85	Broader STEM	Engagement	Yes
Commerce	NOAA Fisheries Education Program	2.24	2.24	2.31	3.46	Broader STEM	Engagement	No
Commerce	Satellite and Information Service	6.36	2.98	3.18	1.11	Broader STEM	Engagement	No
Commerce	NOAA Teacher at Sea Program	0.19	0.60	0.60	0.60	Broader STEM	Pre- & In-Service Educator Performance	No

Agency	Investment	FY 08	FY 09	FY 10	FY 11	Type	Primary Objective	Under-represented Groups
Commerce	NOAA Office of Ocean Exploration and Research (Education Only)	-	0.90	0.90	0.90	Broader STEM	Pre- & In-Service Educator Performance	No
Commerce	National Sea Grant College Program - Education Component	9.73	9.39	9.38	0.75	Broader STEM	Engagement	No
Commerce	Environmental Literacy Grants program	3.83	7.70	10.39	3.93	Broader STEM	Engagement	Yes
Commerce	National Estuarine Research Reserve System	2.82	3.50	4.00	1.99	Broader STEM	Engagement	No
Commerce	Educational Partnership Program with Minority Serving Institutions	13.92	14.98	14.31	14.27	Agency Mission Workforce	Postsecondary STEM Degrees	Yes
Commerce	NIST Summer Institute for Middle School Teachers	0.10	0.20	0.30	0.30	Broader STEM	Pre- & In-Service Educator Performance	No
Commerce	Coral Reef Conservation Program	0.83	0.83	0.84	0.50	Agency Mission Workforce	STEM Careers	Yes
Commerce	Summer Undergraduate Research Fellowship (SURF)	0.51	0.63	0.74	0.60	Agency Mission Workforce	Postsecondary STEM Degrees	No
Commerce	Ernest F. Hollings Undergraduate Scholarship Program	3.97	5.60	5.60	4.59	Agency Mission Workforce	Postsecondary STEM Degrees	Yes
Commerce	Dr. Nancy Foster Scholarship Program	0.47	0.62	0.60	0.51	Agency Mission Workforce	STEM Careers	Yes
Commerce	**Total**	54.51	59.87	62.85	36.37			
Defense	STARBASE	20.00	19.00	20.00	27.00	Broader STEM	Engagement	Yes
Defense	Navy - Historically Black Colleges and Universities/Minority Institutions Research and Education Partnership	1.50	1.50	1.50	1.50	Agency Mission Workforce	STEM Careers	Yes
Defense	University Laboratory Initiative (ULI)	2.60	2.30	2.35	2.30	Agency Mission Workforce	STEM Careers	No

Appendix Table A4. (Continued)

Agency	Investment	FY 08	FY 09	FY 10	FY 11	Type	Primary Objective	Under-represented Groups
Defense	Air Force Consortium Research Fellows Program (CRFP)	-	-	-	0.81	Agency Mission Workforce	Postsecondary STEM Degrees	No
Defense	Army Educational Outreach Program (AEOP)	7.44	7.86	7.74	7.87	Broader STEM	Engagement	Yes
Defense	Iridescent Learning	-	-	0.61	0.86	Broader STEM	Engagement	Yes
Defense	Navy - Science and Engineering Apprenticeship Program (SEAP)	0.27	0.31	0.70	0.78	Broader STEM	Engagement	No
Defense	SeaPerch	0.45	0.75	0.90	1.45	Broader STEM	Engagement	Yes
Defense	The Naval Research Enterprise Intern Program (NREIP)	1.20	1.20	1.90	1.20	Agency Mission Workforce	Postsecondary STEM Degrees	No
Defense	Uniformed Services University of the Health Sciences (USUHS)	0.51	0.53	0.45	0.45	Agency Mission Workforce	STEM Careers	No
Defense	University NanoSatellite Program	1.50	1.60	1.60	1.60	Agency Mission Workforce	Learning	No
Defense	National Defense Education Program (NDEP) K-12 component	1.00	15.00	11.70	11.20	Broader STEM	Engagement	No
Defense	National Defense Education Program (NDEP) Science, Mathematics And Research for Transformation (SMART)	19.00	33.00	31.17	48.80	Agency Mission Workforce	Postsecondary STEM Degrees	No
Defense	National Science Center (NSC)	1.84	1.86	1.98	1.76	Broader STEM	Engagement	No
Defense	National Defense Science and Engineering Graduate (NDSEG) Fellowship Program	33.70	39.34	36.81	37.50	Agency Mission Workforce	Postsecondary STEM Degrees	No
Defense	Awards to Stimulate and Support Undergraduate Research Experiences (ASSURE)	4.50	4.50	4.50	4.50	Agency Mission Workforce	Postsecondary STEM Degrees	Yes
Defense	Stokes Educational Scholarship Program	2.07	1.93	1.94	2.16	Agency Mission Workforce	STEM Careers	Yes

Agency	Investment	FY 08	FY 09	FY 10	FY 11	Type	Primary Objective	Under-represented Groups
Defense	**Total**	97.57	127.68	125.81	151.75			
Education	Graduate Assistance in Areas of National Need (GAANN)	30.00	31.00	31.00	31.00	Broader STEM	Postsecondary STEM Degrees	No
Education	Teacher Loan Forgiveness	15.44	38.63	49.77	66.70	Broader STEM	Pre- & In-Service Educator Performance	No
Education	Minority Science and Engineering Improvement Program	8.58	8.58	9.50	9.48	Broader STEM	Postsecondary STEM Degrees	Yes
Education	Strengthening Predominantly Black Institutions	4.74	8.94	-	5.80	Broader STEM	Institutional Capacity	Yes
Education	Developing Hispanic Serving Institutions STEM and articulation programs	100.00	100.00	100.00	100.00	Broader STEM	Institutional Capacity	Yes
Education	Research, Development, and Dissemination	76.30	52.22	63.81	35.01	Broader STEM	Education R & D	No
Education	Research in Special Education	2.45	10.67	14.64	6.70	Broader STEM	Education R & D	No
Education	High School Longitudinal Study of 2009	6.45	5.14	6.15	6.50	Broader STEM	Education R & D	No
Education	Mathematics and Science Partnerships	179.00	179.00	180.50	175.10	Broader STEM	Pre- & In-Service Educator Performance	Yes
Education	Investing in Innovation	-	-	110.50	74.00	Broader STEM	Education R & D	Yes
Education	Upward Bound Math and Science Program	31.19	35.20	34.87	33.81	Broader STEM	Postsecondary STEM Degrees	Yes
Education	**Total**	**454.15**	**469.38**	**600.75**	**544.11**			
Energy	QuarkNet	0.75	0.75	0.75	0.75	Broader STEM	Pre- & In-Service Educator Performance	No
Energy	National Science Bowl	1.67	1.76	2.45	3.13	Broader STEM	Learning	No
Energy	Minority University Research Associates Program (MURA)	-	-	0.59	0.46	Agency Mission Workforce	STEM Careers	Yes

Appendix Table A4. (Continued)

Agency	Investment	FY 08	FY 09	FY 10	FY 11	Type	Primary Objective	Under-represented Groups
Energy	Plasma/Fusion Science Educator Programs	0.73	0.77	0.77	0.77	Broader STEM	Pre- & In-Service Educator Performance	No
Energy	Advanced Vehicle Competitions	1.39	1.75	2.00	2.00	Agency Mission Workforce	STEM Careers	No
Energy	Graduate Automotive Technology Education	0.50	0.95	1.00	1.00	Agency Mission Workforce	Postsecondary STEM Degrees	No
Energy	Solar Decathlon	2.30	6.40	5.00	5.00	Agency Mission Workforce	Learning	No
Energy	Industrial Assessment Centers	3.40	3.30	6.10	4.10	Agency Mission Workforce	STEM Careers	No
Energy	Visiting Faculty Program (formerly FaST)	0.25	1.54	1.02	0.61	Agency Mission Workforce	STEM Careers	Yes
Energy	Wind for Schools	0.37	0.46	0.63	0.60	Agency Mission Workforce	Postsecondary STEM Degrees	No
Energy	Minority Educational Institution Student Partnership Program	0.55	0.66	1.50	1.50	Agency Mission Workforce	Engagement	Yes
Energy	Special Recruitment Programs/Mickey Leland Fellowship	0.70	0.70	0.70	0.70	Agency Mission Workforce	Engagement	Yes
Energy	HBCU Mathematics, Science & Technology, Engineering and Research Workforce Development Program	-	-	8.97	8.27	Agency Mission Workforce	STEM Careers	Yes
Energy	National Undergraduate Fellowship Program in Plasma Physics and Fusion Energy Sciences	0.37	0.37	0.37	0.37	Agency Mission Workforce	STEM Careers	No

Agency	Investment	FY 08	FY 09	FY 10	FY 11	Type	Primary Objective	Under-represented Groups
Energy	Global Change Education Program	1.47	1.42	1.42	0.88	Agency Mission Workforce	Postsecondary STEM Degrees	No
Energy	American Chemical Society Summer School in Nuclear and Radiochemistry	0.52	0.53	0.55	0.56	Agency Mission Workforce	Learning	No
Energy	Computational Science Graduate Fellowship	6.80	6.80	7.80	6.00	Agency Mission Workforce	Postsecondary STEM Degrees	No
Energy	Office of Science Graduate Fellowship (SCGF) program	-	-	5.00	5.00	Agency Mission Workforce	STEM Careers	No
Energy	Science Undergraduate Laboratory Internships	2.58	2.50	3.80	6.34	Agency Mission Workforce	STEM Careers	No
Energy	Community College Internship	0.32	0.29	0.69	0.66	Agency Mission Workforce	STEM Careers	No
Energy	Hydro Research Fellowships	-	0.90	0.90	2.35	Agency Mission Workforce	STEM Careers	No
Energy	**Total**	24.66	31.86	52.00	51.05			
Environmental Protection Agency	Environmental Education Grants	3.40	3.40	2.20	3.70	Broader STEM	Learning	No
Environmental Protection Agency	National Environmental Education Training Program	2.00	2.00	2.20	2.40	Broader STEM	Pre- & In-Service Educator Performance	Yes
Environmental Protection Agency	P3-People, Prosperity & the Planet-Award: A National Student Design Competition for Sustainability	1.30	1.20	2.00	1.60	Broader STEM	Engagement	No

Appendix Table A4. (Continued)

Agency	Investment	FY 08	FY 09	FY 10	FY 11	Type	Primary Objective	Under-represented Groups
Environmental Protection Agency	Cooperative Training Partnership in Environmental Sciences Research	2.00	1.50	1.50	0.50	Agency Mission Workforce	STEM Careers	No
Environmental Protection Agency	Greater Research Opportunities (GRO) Fellowships for Undergraduate Environmental Study	0.60	1.30	1.50	2.00	Agency Mission Workforce	Postsecondary STEM Degrees	No
Environmental Protection Agency	Science to Achieve Results Graduate Fellowship Program	8.22	4.24	6.39	9.50	Agency Mission Workforce	Postsecondary STEM Degrees	No
Environmental Protection Agency	University of Cincinnati/EPA Research Training Grant	0.60	0.60	0.60	0.60	Agency Mission Workforce	STEM Careers	No
	Total	**18.12**	**14.24**	**16.39**	**20.30**			
Health & Human Services	Science Education Drug Abuse Partnership Award	2.16	2.46	2.30	1.90	Broader STEM	Learning	Yes
Health & Human Services	NIAID Science Education Awards	0.35	0.71	1.06	0.61	Broader STEM	Learning	No
Health & Human Services	NIH Science Education Partnership Award (SEPA)	16.18	22.21	18.32	18.90	Broader STEM	Learning	Yes
Health & Human Services	Curriculum Supplement Series	0.76	0.36	0.34	0.35	Broader STEM	Learning	No
Health & Human Services	Mathematics and Science Cognition and Learning (MSCL) Program	4.70	10.10	10.40	9.20	Broader STEM	Education R & D	No
Health & Human Services	Office of Science Education K-12 Program	2.10	2.11	2.27	2.27	Broader STEM	Engagement	Yes
Health & Human Services	NLM Institutional Grants for Research Training in Biomedical Informatics	10.33	14.66	10.14	12.20	Agency Mission Workforce	Postsecondary STEM Degrees	Yes

Agency	Investment	FY 08	FY 09	FY 10	FY 11	Type	Primary Objective	Under-represented Groups
Health & Human Services	Research Supplements to Promote Diversity in Health-Related Research	70.56	83.43	68.98	66.20	Agency Mission Workforce	Postsecondary STEM Degrees	Yes
Health & Human Services	Ruth L. Kirschstein National Research Service Award Institutional Research Training Grants (T32, T35)	259.08	266.47	230.84	219.40	Agency Mission Workforce	STEM Careers	Yes
Health & Human Services	Ruth L. Kirschstein NRSA for Individual Predoctoral Fellows, including Underrepresented Racial/Ethnic Groups, Students from Disadvantaged Backgrounds, and Predoctoral Students with Disabilities	47.57	55.55	56.88	54.10	Agency Mission Workforce	STEM Careers	Yes
Health & Human Services	MARC U-STAR NRSA Program	16.76	21.25	21.25	20.90	Agency Mission Workforce	Postsecondary STEM Degrees	Yes
Health & Human Services	Short Courses on Mathematical, Statistical, and Computational Tools for Studying Biological Systems	0.32	0.33	0.70	0.90	Agency Mission Workforce	Learning	No
Health & Human Services	Short Courses in Integrative and Organ Systems Pharmacology	0.68	0.75	0.67	0.67	Agency Mission Workforce	Learning	No
Health & Human Services	Initiative for Maximizing Student Development	16.44	22.34	21.41	22.30	Agency Mission Workforce	Postsecondary STEM Degrees	Yes
Health & Human Services	RISE (Research Initiative for Scientific Enhancement)	18.57	25.69	24.44	26.20	Agency Mission Workforce	Postsecondary STEM Degrees	Yes
Health & Human Services	Bridges to the Baccalaureate Program	4.00	7.26	6.46	6.71	Agency Mission Workforce	Postsecondary STEM Degrees	Yes
Health & Human Services	Bridges to the Doctorate	1.29	2.18	2.98	3.17	Agency Mission Workforce	Postsecondary STEM Degrees	Yes

Appendix Table A4. (Continued)

Agency	Investment	FY 08	FY 09	FY 10	FY 11	Type	Primary Objective	Under-represented Groups
Health & Human Services	Postbaccalaureate Research Education Program (PREP)	3.03	6.73	5.78	7.70	Agency Mission Workforce	Postsecondary STEM Degrees	Yes
Health & Human Services	Cancer Education Grants Program	6.75	7.31	6.76	5.70	Agency Mission Workforce	Learning	No
Health & Human Services	National Cancer Institute Cancer Education and Career Development Program	16.21	18.99	20.44	18.40	Agency Mission Workforce	Learning	No
Health & Human Services	Summer Institute for Training in Biostatistics	-	1.45	1.45	1.50	Agency Mission Workforce	Postsecondary STEM Degrees	No
Health & Human Services	Short-Term Research Education Program to Increase Diversity in Health-Related Research	2.05	3.17	4.19	4.00	Agency Mission Workforce	Postsecondary STEM Degrees	Yes
Health & Human Services	Short Term Educational Experiences for Research (STEER) in the Environmental health Sciences for Undergraduates and High School Students	0.57	0.76	0.57	0.50	Agency Mission Workforce	Engagement	No
Health & Human Services	NINDS Diversity Research Education Grants in Neuroscience	0.25	0.71	0.82	1.00	Agency Mission Workforce	Postsecondary STEM Degrees	Yes
Health & Human Services	Short Courses in Population Research (Education Programs for Population Research R25)	0.70	0.92	0.75	0.80	Agency Mission Workforce	Learning	No
Health & Human Services	Graduate Program Partnerships	17.60	16.60	16.70	16.00	Agency Mission Workforce	STEM Careers	No
Health & Human Services	Post-baccalaureate Intramural Research Training Award Program	21.54	21.48	24.81	23.50	Agency Mission Workforce	Postsecondary STEM Degrees	No
Health & Human Services	Student Intramural Research Training Award Program	5.78	5.12	5.87	5.40	Agency Mission Workforce	Learning	No

Agency	Investment	FY 08	FY 09	FY 10	FY 11	Type	Primary Objective	Under-represented Groups
Health & Human Services	Technical Intramural Research Training Award	2.14	2.14	2.24	2.24	Agency Mission Workforce	STEM Careers	No
Health & Human Services	Undergraduate Scholarship Program for Individuals from Disadvantaged Backgrounds	2.20	2.30	2.40	2.50	Agency Mission Workforce	Postsecondary STEM Degrees	Yes
Health & Human Services	Clinical Research Training Program	1.00	1.00	1.10	0.85	Agency Mission Workforce	Learning	No
Health & Human Services	Health Careers Opportunity Program	-	-	-	22.00	Broader STEM	Postsecondary STEM Degrees	Yes
Health & Human Services	Public Health Traineeship	-	-	-	1.40	Agency Mission Workforce	STEM Careers	No
	Total	**551.66**	**626.52**	**573.32**	**579.45**			
Homeland Security	National Nuclear Forensics Expertise Development Program	-	-	-	2.50	Agency Mission Workforce	STEM Careers	Yes
Homeland Security	Homeland Security STEM Career Development Grant Program	3.50	2.50	2.50	1.00	Agency Mission Workforce	STEM Careers	No
Homeland Security	Scientific Leadership Awards Program	3.80	3.80	3.40	1.00	Agency Mission Workforce	STEM Careers	Yes
Homeland Security	**Total**	**7.30**	**6.30**	**5.90**	**4.50**			
Interior	Educational Component of the National Geologic Mapping Program (EDMAP)	0.49	0.52	0.57	0.57	Agency Mission Workforce	Postsecondary STEM Degrees	No
Interior	Conservation and Land Management Internship Program	-	-	-	1.74	Agency Mission Workforce	STEM Careers	Yes

Appendix Table A4. (Continued)

Agency	Investment	FY 08	FY 09	FY 10	FY 11	Type	Primary Objective	Under-represented Groups
Interior	George Melendez Wright Climate Change Youth Initiative	-	-	-	0.91	Agency Mission Workforce	STEM Careers	Yes
Interior	Geoscientists-in-the-Parks Program	-	-	-	0.33	Agency Mission Workforce	Learning	Yes
Interior	**Total**	**0.49**	**0.52**	**0.57**	**3.55**			
NASA	LDCM	0.14	0.54	0.30	0.68	Broader STEM	Pre- & In-Service Educator Performance	Yes
NASA	HST	1.42	1.35	1.25	1.29	Broader STEM	Learning	Yes
NASA	Cassini	1.55	1.70	1.65	1.26	Broader STEM	Learning	Yes
NASA	Aqua	-	0.48	0.43	0.44	Broader STEM	Engagement	No
NASA	GCCE - Global Climate Change Education	7.00	10.00	10.00	3.50	Broader STEM	Learning	Yes
NASA	Astrophysics Forum	-	0.99	1.00	1.02	Broader STEM	Institutional Capacity	Yes
NASA	Heliophysics	-	0.82	0.73	0.79	Broader STEM	Institutional Capacity	Yes
NASA	Planetary Science E/PO Forum	-	0.92	0.89	0.87	Broader STEM	Institutional Capacity	Yes
NASA	Earth Science E/PO Forum	-	0.76	0.87	0.89	Broader STEM	Institutional Capacity	Yes
NASA	Chandra	1.92	1.85	1.82	1.77	Broader STEM	Engagement	No
NASA	LTP - Learning Technologies Project	1.28	0.84	0.71	0.54	Broader STEM	Education R & D	No
NASA	EPOESS	0.79	4.58	6.91	7.00	Broader STEM	Learning	Yes
NASA	Juno	0.58	1.21	1.31	1.27	Broader STEM	Learning	Yes
NASA	DAWN	0.30	0.27	0.36	0.53	Broader STEM	Learning	Yes
NASA	GRAIL	0.20	0.31	0.41	0.53	Broader STEM	Engagement	Yes
NASA	NAS - NASA Aerospace Scholars			0.30	0.30	Broader STEM	Engagement	No
NASA	NES - NASA Explorer Schools	8.09	4.31	4.99	3.10	Broader STEM	Engagement	No
NASA	NASA's Beginning Engineering, Science and Technology (BEST) students	-	-	-	0.43	Broader STEM	Education R & D	Yes

Agency	Investment	FY 08	FY 09	FY 10	FY 11	Type	Primary Objective	Under-represented Groups
NASA	SOFIA (Stratospheric Observatory for Infrared Astronomy) Education and Public Outreach	0.24	0.36	0.60	0.61	Broader STEM	Pre- & In-Service Educator Performance	Yes
NASA	NSTI-MI - NASA Science and Technology Institute for Minority Institutions	2.00	1.96	2.46	2.11	Agency Mission Workforce	Institutional Capacity	Yes
NASA	Space Grant - National Space Grant College and Fellowship Program	39.71	38.30	44.50	45.50	Agency Mission Workforce	Postsecondary STEM Degrees	Yes
NASA	eEducation Small Projects/Central Operation of Resources for Educators (CORE)	0.60	0.49	0.40	0.49	Broader STEM	Pre- & In-Service Educator Performance	Yes
NASA	NETS - NASA Education Technologies Services	1.40	1.30	1.00	0.47	Broader STEM	Learning	Yes
NASA	Curriculum Improvement Partnership Award for the Integration of Research into the Undergraduate Curriculum (CIPAIR)	2.75	2.71	3.11	0.60	Agency Mission Workforce	Postsecondary STEM Degrees	Yes
NASA	University student launch initiative	-	-	-	0.32	Agency Mission Workforce	Engagement	Yes
NASA	Aeronautics Scholarship	1.80	1.80	1.80	1.60	Agency Mission Workforce	Postsecondary STEM Degrees	No
NASA	Innovation in Aeronautics Instruction Competition	1.10	1.10	1.10	0.46	Agency Mission Workforce	Institutional Capacity	No
NASA	JPFP - Jenkins Pre-Doctoral Fellowship Program	2.56	2.53	2.63	3.40	Agency Mission Workforce	Postsecondary STEM Degrees	Yes
NASA	USRP - Undergraduate Student Research Project	4.00	3.48	2.97	3.30	Agency Mission Workforce	Postsecondary STEM Degrees	No
NASA	EFP - Education Flight Projects	1.20	3.11	2.99	2.70	Broader STEM	Engagement	No
NASA	SEMAA - Science Engineering Mathematics and Aerospace Academy	2.51	1.91	3.09	2.13	Broader STEM	Learning	Yes

Appendix Table A4. (Continued)

Agency	Investment	FY 08	FY 09	FY 10	FY 11	Type	Primary Objective	Under-represented Groups
NASA	SEED - Systems Engineering Educational Discovery	0.29	0.37	0.41	0.30	Agency Mission Workforce	STEM Careers	No
NASA	Reduced Gravity Student Flight Opportunity Project	-	0.36	0.36	0.30	Agency Mission Workforce	Learning	No
NASA	SOI - Summer of Innovation	-	-	10.00	10.10	Broader STEM	Engagement	Yes
NASA	HEOMD Space Grant Project	1.36	1.55	1.03	1.07	Agency Mission Workforce	Postsecondary STEM Degrees	Yes
NASA	NSBRI Higher Education Activities - National Space Biomedical Research Institute	0.72	0.74	0.75	0.65	Agency Mission Workforce	STEM Careers	Yes
NASA	MESSENGER	0.43	0.36	0.30	0.34	Broader STEM	Learning	No
NASA	URC - University Research Centers	13.93	14.57	14.06	10.40	Agency Mission Workforce	Postsecondary STEM Degrees	Yes
NASA	Aura	-	0.38	0.37	0.48	Broader STEM	Learning	No
NASA	GSRP - Graduate Student Researchers Program	5.20	4.30	4.40	3.30	Agency Mission Workforce	STEM Careers	No
NASA	Innovation in Higher Education STEM Education	-	-	0.96	4.70	Agency Mission Workforce	Postsecondary STEM Degrees	No
NASA	AESP - Aerospace Education Services Project	4.90	5.50	2.50	3.90	Broader STEM	Pre- & In-Service Educator Performance	Yes
NASA	Mars E/PO Informal Ed	0.99	0.77	0.81	1.00	Broader STEM	Engagement	No
NASA	Mars E/PO Formal Ed	1.20	1.00	1.30	1.10	Broader STEM	Learning	No
NASA	GLOBE (Learning and Observations to Benefit the Environment) Program		4.40	3.00	5.00	Broader STEM	Engagement	No
NASA	TCUP - NASA Tribal College and University Project	1.62	1.68	1.59	1.50	Agency Mission Workforce	Postsecondary STEM Degrees	Yes

Agency	Investment	FY 08	FY 09	FY 10	FY 11	Type	Primary Objective	Under-represented Groups
NASA	LEARN - Learning Environment and Research Network	2.40	3.00	3.00	2.70	Broader STEM	Engagement	No
NASA	SIMulation-Based Engineering and Science Teacher Professional Development	-	-	0.39	0.40	Broader STEM	Pre- & In-Service Educator Performance	No
NASA	INSPIRE - Interdisciplinary National Science Program Incorporating Research and Education Experience	2.85	3.42	2.52	3.23	Broader STEM	Engagement	Yes
NASA	LERCIP - Lewis Educational Research Collaborative Internship Project (College)	0.76	0.97	0.90	0.60	Agency Mission Workforce	Postsecondary STEM Degrees	No
NASA	MSP - MUREP Small Projects	1.50	1.80	1.70	1.80	Agency Mission Workforce	Institutional Capacity	Yes
NASA	LARSS - NASA Langley Aerospace Research Student Scholars Program	1.00	1.10	1.30	1.30	Agency Mission Workforce	Postsecondary STEM Degrees	No
NASA	MUST - Motivating Undergraduates in Science and Technology	1.90	1.90	2.40	1.80	Agency Mission Workforce	Postsecondary STEM Degrees	Yes
NASA	CEP - Career Exploration Project	1.16	1.15	1.30	1.20	Agency Mission Workforce	Engagement	Yes
NASA	Space technology research fellowships	-	-	-	7.00	Agency Mission Workforce	Learning	No
NASA	**Total**	**125.33**	**139.27**	**155.91**	**154.07**			
National Science Foundation	Opportunities for Enhancing Diversity in the Geosciences	4.57	11.79	4.18	3.60	Broader STEM	Engagement	Yes
National Science Foundation	Geoscience Education	1.63	2.74	2.02	1.50	Broader STEM	Learning	Yes
National Science Foundation	Climate Change Education (CCE)	-	9.95	10.24	10.00	Broader STEM	Learning	Yes

Appendix Table A4. (Continued)

Agency	Investment	FY 08	FY 09	FY 10	FY 11	Type	Primary Objective	Under-represented Groups
National Science Foundation	CITEAM	9.90	0.00	4.85	5.00	Broader STEM	Learning	Yes
National Science Foundation	CISE Pathways to Revitalized Undergraduate Computing Education (CPATH)	5.00	5.00	4.37	1.50	Broader STEM	Institutional Capacity	Yes
National Science Foundation	Geoscience Teacher Training (GEO-Teach)	3.00	3.00	2.98	2.00	Broader STEM	Pre- & In-Service Educator Performance	Yes
National Science Foundation	Broadening Participation in Computing (BPC)	14.00	14.00	14.00	14.00	Broader STEM	Postsecondary STEM Degrees	Yes
National Science Foundation	Global Learning and Observations to Benefit the Environment (GLOBE)	1.10	1.12	1.10	1.10	Broader STEM	Engagement	No
National Science Foundation	Undergraduate Research and Mentoring in the Biological Sciences (URM)	5.09	4.68	9.00	3.00	Broader STEM	Engagement	Yes
National Science Foundation	Transforming Undergraduate Biology Education (TUBE)	-	0.90	5.06	15.00	Broader STEM	Engagement	Yes
National Science Foundation	Computing Education for the 21st Century (CE21)	-	-	-	5.50	Broader STEM	Learning	Yes
National Science Foundation	Graduate STEM Fellows in K-12 Education (GK-12)	54.60	58.84	55.97	52.95	Broader STEM	Learning	Yes
National Science Foundation	Math and Science Partnership (MSP)	47.87	85.99	57.93	57.08	Broader STEM	Education R & D	Yes
National Science Foundation	Robert Noyce Scholarship (Noyce) Program	55.05	115.00	54.93	54.89	Broader STEM	Pre- & In-Service Educator Performance	Yes
National Science Foundation	Science, Technology, Engineering, and Mathematics Talent Expansion Program (STEP)	29.48	29.09	31.64	33.44	Broader STEM	Postsecondary STEM Degrees	Yes
National Science Foundation	Enhancing the Mathematical Sciences Workforce in the 21st Century (EMSW21)	19.46	26.95	15.07	16.47	Broader STEM	Postsecondary STEM Degrees	Yes
National Science Foundation	Interdisciplinary Training for Undergraduates in Biological and Mathematical Sciences (UBM)	2.32	2.71	2.70	2.10	Broader STEM	Postsecondary STEM Degrees	Yes

Agency	Investment	FY 08	FY 09	FY 10	FY 11	Type	Primary Objective	Under-represented Groups
National Science Foundation	International Research Experiences for Students (IRES)	2.71	4.21	3.43	3.15	Broader STEM	Postsecondary STEM Degrees	Yes
National Science Foundation	Scholarships in Science, Technology, Engineering, and Mathematics (S-STEM)	92.40	61.22	75.96	75.00	Broader STEM	Postsecondary STEM Degrees	Yes
National Science Foundation	Centers for Ocean Sciences Education Excellence	5.24	7.19	5.70	5.24	Broader STEM	Institutional Capacity	No
National Science Foundation	Advanced Technological Education (ATE)	51.46	51.85	64.51	64.00	Broader STEM	Postsecondary STEM Degrees	Yes
National Science Foundation	Research and Evaluation on Education in Science and Engineering (REESE)	41.66	42.60	45.67	45.59	Broader STEM	Education R & D	No
National Science Foundation	Research in Disabilities Education (RDE)	5.93	6.88	6.92	6.48	Broader STEM	Postsecondary STEM Degrees	Yes
National Science Foundation	Informal Science Education (ISE)	64.45	65.72	65.85	64.22	Broader STEM	Education R & D	Yes
National Science Foundation	Louis Stokes Alliances for Minority Participation (LSAMP)	40.47	42.50	44.55	45.63	Broader STEM	Postsecondary STEM Degrees	Yes
National Science Foundation	Transforming Undergrad Education in STEM (TUES)	37.28	40.86	41.60	40.92	Broader STEM	Learning	Yes
National Science Foundation	Excellence Awards in Science and Engineering (EASE)	5.57	5.15	5.18	5.18	Broader STEM	Engagement	Yes
National Science Foundation	Research Experiences for Undergraduates (REU)	62.67	100.47	80.99	63.55	Broader STEM	STEM Careers	Yes
National Science Foundation	Nanotechnology Undergraduate Education in Engineering	1.08	2.00	1.83	1.50	Broader STEM	Learning	Yes
National Science Foundation	Research Experiences for Teachers (RET) in Engineering and Computer Science	3.97	5.79	5.41	4.00	Broader STEM	Pre- & In-Service Educator Performance	Yes
National Science Foundation	Engineering Education (EE)	11.50	22.89	13.74	10.85	Broader STEM	Education R & D	Yes
National Science Foundation	Alliances for Graduate Education and the Professoriate (AGEP)	15.85	17.18	16.73	16.70	Broader STEM	Education R & D	Yes
National Science Foundation	Tribal Colleges and Universities Program (TCUP)	12.80	13.39	13.35	13.31	Broader STEM	Institutional Capacity	Yes

Appendix Table A4. (Continued)

Agency	Investment	FY 08	FY 09	FY 10	FY 11	Type	Primary Objective	Under-represented Groups
National Science Foundation	Historically Black Colleges and Universities Undergraduate Program (HBCU-UP)	29.74	31.13	32.06	31.94	Broader STEM	Institutional Capacity	Yes
National Science Foundation	Graduate Research Fellowship Program (GRFP)	96.02	162.44	136.13	137.68	Broader STEM	Postsecondary STEM Degrees	Yes
National Science Foundation	Integrative Graduate Education and Research Traineeship (IGERT) Program	64.76	77.99	69.70	60.76	Broader STEM	Postsecondary STEM Degrees	Yes
National Science Foundation	East Asia & Pacific Summer Institutes for U.S. Graduate Students (EAPSI)	1.75	1.52	1.74	2.40	Broader STEM	Engagement	Yes
National Science Foundation	Federal Cyber Service: Scholarship for Service (SFS)	11.37	14.88	14.87	14.96	Agency Mission Workforce	STEM Careers	Yes
National Science Foundation	Discovery Research K-12 (DR-K12)	99.25	108.41	118.38	119.90	Broader STEM	Education R & D	Yes
National Science Foundation	Innovative Technology Experiences for Students and Teachers (ITEST)	28.72	27.86	20.85	25.00	Broader STEM	Engagement	Yes
National Science Foundation	Research on Gender in Science and Engineering (GSE)	10.13	11.40	11.57	10.47	Broader STEM	Education R & D	Yes
National Science Foundation	**Total**	**1049.85**	**1297.29**	**1172.76**	**1147.57**			
Nuclear Regulatory Commission	Nuclear Education Curriculum Development Program	4.72	4.72	4.70	4.70	Agency Mission Workforce	Institutional Capacity	No
Nuclear Regulatory Commission	Minority Serving Institutions Program (MSIP)	1.00	1.42	2.84	0.70	Agency Mission Workforce	Postsecondary STEM Degrees	Yes
Nuclear Regulatory Commission	Integrated University Program - Scholarship and Fellowship Program	-	8.40	15.00	5.00	Agency Mission Workforce	Postsecondary STEM Degrees	No
	Total	**5.72**	**14.54**	**22.54**	**10.40**			

Agency	Investment	FY 08	FY 09	FY 10	FY 11	Type	Primary Objective	Under-represented Groups
Transportation	Air Transportation Centers of Excellence (COE)	13.20	14.10	16.40	12.50	Agency Mission Workforce	Postsecondary STEM Degrees	No
Transportation	University Transportation Centers (UTC) Program	74.44	83.45	83.67	80.00	Agency Mission Workforce	Postsecondary STEM Degrees	No
Transportation	Dwight David Eisenhower Transportation Fellowship Program	1.96	1.99	2.01	2.00	Agency Mission Workforce	Postsecondary STEM Degrees	No
Transportation	Garrett A. Morgan Technology and Transportation Education Program (GAM)	1.11	1.13	1.14	1.10	Agency Mission Workforce	Learning	Yes
Transportation	Summer Transportation Institute Program for Diverse Groups (STIPDG)	0.64	0.65	0.65	3.90	Agency Mission Workforce	Engagement	Yes
Transportation	National Summer Transportation Institute Program (STI)	-	-	-	0.80	Agency Mission Workforce	Engagement	Yes
Transportation	**Total**	**91.36**	**101.32**	**103.87**	**100.30**			
	Grand Total	**2561.48**	**2975.60**	**2983.56**	**2891.34**			

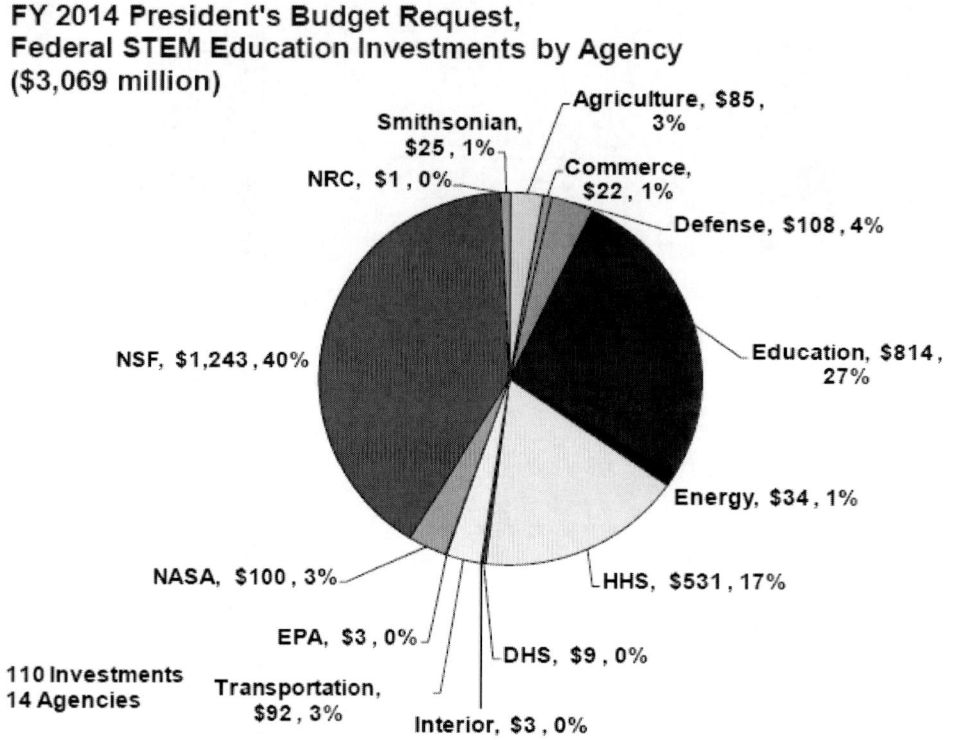

Figure A4. FY 2014 President's Budget Request, Federal STEM Education Investments by Agency.

Table A5. Federal Stem Education Funding by Agency in Millions FY 2012 and2014

	2012 Enacted	2014 Requested
Agriculture	$ 88	$ 85
Commerce	$ 40	$ 22
Defense	$ 153	$ 108
DHS	$ 11	$ 9
Education	$ 529	$ 814
Energy	$ 47	$ 34
EPA	$ 26	$ 3
HHS	$ 576	$ 531
Interior	$ 3	$ 3
NASA	$ 149	$ 100
NRC	$ 16	$ 1
NSF	$ 1,154	$1,243
Smithsonian	$ 0	$ 25
Transportation	$ 99	$ 92
Total	**$ 2,891**	**$ 3,069**

Appendix Table A6.
STEM Education Funding in Millions by Agency[3]

Agency	Sub-Agency	Investment	FY 12 Enacted	FY 14 Requested
Agriculture	NIFA	1890 Facilities Grant Program	19.7	19.7
Agriculture	NIFA	1890 Institutions Capacity Building Grants Program: Extension	6.4	6.4
Agriculture	NIFA	1890 Institutions Capacity Building Grants Program: Teaching	6.4	6.4
Agriculture	NIFA	4-H Science, 4-H Youth Development Program	24.0	24.0
Agriculture	APHIS	AgDiscovery	0.5	0.5
Agriculture	NIFA	Agriculture in the Classroom	0.6	-
Agriculture	NIFA	AITC Secondary Postsecondary Agriculture Education Challenge Grants (SPECA)	0.9	-
Agriculture	NIFA	Alaska Native-Serving and Native Hawaiian-Serving Institutions Education Competitive Grants Program	3.2	3.2
Agriculture	NIFA	Distance Education Grants for Institutions of Higher Education in Insular Areas (DEG)	0.8	-

[3] Only investments with a budget of $300,000 or greater in at least one year from FY 2012 to FY 2014 are included in the table. Basic information on investments with a budget consistently below $300,000 was and will continue to be collected as part of the annual NSTC Federal STEM Education Inventory. Investments that appear in the FY 2011 inventory (Appendix Table A4) and do not appear in the table were either eliminated in FY 2012 or fell below $300,000 a year.

Appendix Table A6. (Continued)

Agency	Sub-Agency	Investment	FY 12 Enacted	FY 14 Requested
Agriculture	NIFA	Food and Agricultural Sciences National Needs Graduate and Postgraduate Fellowship Grant Program	3.2	-
Agriculture	NIFA	Higher Education Challenge Grants (HEC)	4.8	-
Agriculture	NIFA	Higher Education Multicultural Scholars Program (MSP)	1.0	-
Agriculture	NIFA	Hispanic Serving Institutions Education Grants Program	9.2	9.2
Agriculture	NIFA	Insular Programs	-	1.7
Agriculture	NIFA	NIFA Fellowship Grants Program	6.1	13.7
Agriculture	NIFA	Resident Instruction Grants Program for Institutions of Higher Education in Insular Areas	0.9	-
Agriculture	NIFA	Women and Minorities in Science, Technology, Engineering and Mathematics Fields Program (WAMS)	0.4	-
Agriculture Total			**88.1**	**84.8**

Agency	Sub-Agency	Investment	FY 12 Enacted	FY 14 Requested
Commerce	NOAA	Competitive Education Grants (including Environmental Literacy Grants)	5.1	-
Commerce	NOAA	Coral Reef Conservation Program	0.5	-
Commerce	NOAA	Dr. Nancy Foster Scholarship Program	0.5	-
Commerce	NOAA	Educational Partnership Program with Minority Serving Institutions	12.5	14.4
Commerce	NOAA	Ernest F. Hollings Undergraduate Scholarship Program	4.9	5.4
Commerce	NOAA	National Estuarine Research Reserve System	0.6	-
Commerce	NOAA	National Sea Grant College Program - Education	0.8	-
Commerce	NIST	NIST Summer Institute for Middle School Teachers	0.3	-
Commerce	NOAA	NOAA Bay Watershed Education and Training (B-WET)	5.5	-

Appendix Table A6. (Continued)

Agency	Sub-Agency	Investment	FY 12 Enacted	FY 14 Requested
Commerce	NOAA	NOAA Fisheries Education Program*	3.5	-
Commerce	NOAA	NOAA Office of Ocean Exploration and Research (Education Only)	0.9	-
Commerce	NOAA	NOAA Teacher at Sea Program	0.6	-
Commerce	NOAA	Satellite and Information Service*	3.2	-
Commerce	NIST	STEM Pipeline for the Next Generation Scientists and Engineers.	0.4	1.0
Commerce	NIST	Summer Undergraduate Research Fellowship (SURF)	0.8	0.8
		*FY 2012 figures are estimates.		
Commerce Total			**40.0**	**21.6**
Defense		Army Educational Outreach Program (AEOP)	8.2	9.4
Defense		Awards to Stimulate and Support Undergraduate Research Experiences (ASSURE)	4.5	4.5

Agency	Sub-Agency	Investment	FY 12 Enacted	FY 14 Requested
Defense		DoD STARBASE Program	25.0	-
Defense		Navy Historically Black Colleges and Universities/Minority Institutions Research and Education Partnership	1.4	1.5
Defense		Iridescent Learning	2.5	-
Defense		National Defense Education Program (NDEP) K-12 component	16.6	-
Defense		National Defense Education Program (NDEP) Science, Mathematics And Research for Transformation (SMART)	43.3	48.7
Defense		National Defense Science and Engineering Graduate (NDSEG) Fellowship Program	39.7	38.0
Defense		National Science Center (NSC)	1.9	-
Defense		Navy - Science and Engineering Apprenticeship Program (SEAP)	0.8	0.7
Defense		SeaPerch	1.5	1.5

Appendix Table A6. (Continued)

Agency	Sub-Agency	Investment	FY 12 Enacted	FY 14 Requested
Defense	NSA	Stokes Educational Scholarship Program	2.0	1.8
Defense		The Naval Research Enterprise Intern Program (NREIP)	1.3	0.9
Defense		Uniformed Services University of the Health Sciences (USUHS)	0.5	-
Defense		University Laboratory Initiative (ULI)	2.2	-
Defense		University NanoSatellite Program	1.2	1.2
Defense Total			**152.6**	**108.2**
Education	OPE	Developing Hispanic Serving Institutions STEM and articulation programs	100.0	100.0
Education	OII	Fund for the Improvement of Education (FIE)	-	29.7
Education	OPE	Graduate Assistance in Areas of National Need (GAANN)	30.9	30.9

Agency	Sub-Agency	Investment	FY 12 Enacted	FY 14 Requested
Education	IES	High School Longitudinal Study of 2009	6.7	3.0
Education	OESE	Improving Teacher Quality State Grants/Effective Teacher and Leader State Grants	-	-
Education	OII	Investing in Innovation	28.5	45.0
Education	OESE	Mathematics and Science Partnerships/Effective Teaching and Learning for a Complete Education	149.7	149.7
Education	OPE	Minority Science and Engineering Improvement Program	9.5	9.5
Education	IES	Research in Special Education	3.3	6.5
Education	IES	Research, Development, and Dissemination	31.2	45.0
Education	OESE	STEM Innovation Networks	-	265.0
Education	OPE	Strengthening Predominantly Black Institutions	5.7	5.7

Appendix Table A6. (Continued)

Agency	Sub-Agency	Investment	FY 12 Enacted	FY 14 Requested
Education	OESE	Teacher Incentive Fund	40.3	-
Education	OPE	Teacher Loan Forgiveness	79.2	80.0
Education	OPE	Upward Bound Math and Science Program	43.8	43.8
Education Total			**528.7**	**813.8**
Energy	Office of Energy Efficiency and Renewable Energy, Vehicle Technologies	Advanced Vehicle Competitions	2.0	2.0
Energy	Office of Science, Office of Nuclear Physics and Office of Basic Energy Sciences	American Chemical Society Summer School in Nuclear and Radiochemistry	0.6	-
Energy	Office of Science, Office of Workforce Development for Teachers & Scientists	Community College Internship (formerly Community College Institute of Science and Technology)	0.6	0.7
Energy	Office of Science, Advanced Scientific Computing Research	Computational Sciences Graduate Fellowship	6.0	-

Agency	Sub-Agency	Investment	FY 12 Enacted	FY 14 Requested
Energy	Office of Science, Office of Workforce Development for Teachers & Scientists	Visiting Faculty Program (formerly Faculty and Student Teams)	1.2	1.3
Energy	Office of Science, Office of Biological and Environmental Research	Global Change Education Program	0.6	-
Energy	Office of Energy Efficiency and Renewable Energy, Vehicle Technologies	Graduate Automotive Technology Education	1.0	-
Energy	Office of Environmental Management	HBCU Mathematics, Science & Technology, Engineering and Research Workforce Development Program	8.3	8.0
Energy	Office of Energy Efficiency and Renewable Energy, Industrial Technologies	Industrial Assessment Centers	3.7	6.0
Energy	Office of Economic Impact and Diversity	Minority Educational Institution Student Partnership Program	1.2	1.1
Energy	Office of Energy Efficiency and Renewable Energy, Solar Energy Technologies	Minority University Research Associates Program (MURA)	0.5	0.5

Appendix Table A6. (Continued)

Agency	Sub-Agency	Investment	FY 12 Enacted	FY 14 Requested
Energy	Office of Science, Office of Workforce Development for Teachers and Scientists	National Science Bowl	2.7	2.8
Energy	Office of Science, Office of Fusion Energy Sciences	National Undergraduate Fellowship Program in Plasma Physics and Fusion Energy Sciences	0.4	-
Energy	Office of Science, Office of Workforce Development for Teachers and Scientists	Graduate Student Research Program (formerly Office of Science Graduate Fellowship)	5.0	2.0
Energy	Office of Science, Office of Fusion Energy Sciences	Plasma/Fusion Science Educator Programs	0.8	-
Energy	Office of Science, High Energy Physics	QuarkNet	0.6	-
Energy	Office of Science, Office of Workforce Development for Teachers and Scientists	Science Undergraduate Laboratory Internships	6.5	7.3
Energy	Office of Energy Efficiency and Renewable Energy, Building Technologies	Solar Decathlon	4.2	2.0
Energy	Office of Fossil Energy	Special Recuitment Programs/Mickey Leland Fellowship	0.7	0.7

Agency	Sub-Agency	Investment	FY 12 Enacted	FY 14 Requested
Energy	Office of Energy Efficiency and Renewable Energy, Wind Energy	Wind for Schools	0.9	-
Energy Total			**47.5**	**34.4**
EPA	ORD	Cooperative Training Partnership in Environmental Sciences Research	0.5	0.2
EPA	Office of Environmental Education	Environmental Education Grants	3.5	-
EPA	ORD	Greater Research Opportunities (GRO) Fellowships for Undergraduate Environmental Study	2.1	-
EPA	Office of Environmental Education	National Environmental Education and Training Partnership	2.0	-
EPA	ORD	P3-People, Prosperity & the Planet-Award: A National Student Design Competition for Sustainability	3.1	2.6
EPA	NCER	Science to Achieve Results Graduate Fellowship Program	14.0	-
EPA	ORD	University of Cincinnati/EPA Research Training Grant	0.6	0.6

Appendix Table A6. (Continued)

Agency	Sub-Agency	Investment	FY 12 Enacted	FY 14 Requested
EPA Total			**25.8**	**3.4**
HHS	NIH, NIGMS	Bridges to the Baccalaureate Program	6.3	7.2
HHS	NIH, Intramural Training	Clinical Research Training Program	0.2	-
HHS	NIH, Office of Science Education	Curriculum Supplement Series	0.3	-
HHS	HRSA	Health Careers Opportunity Program	15.0	-
HHS	NIH, NIGMS	Initiative for Maximizing Student Development	2.7	2.2
HHS	NIH, NIGMS	MARC U-STAR NRSA Program	18.2	18.0
HHS	NIH, NICHD	Mathematics and Science Cognition and Learning (MSCL) Program	3.7	3.9
HHS		Medical Research Scholars Program (MRSP)	-	0.2

Agency	Sub-Agency	Investment	FY 12 Enacted	FY 14 Requested
HHS	NIH, NCI	National Cancer Institute Cancer Education and Career Development Program	0.9	1.1
HHS	NIH, NIAID	Science Education Awards	1.1	-
HHS	NIH, NINDS	Diversity Research Education Grants in Neuroscience	2.7	-
HHS	NIH, NLM	Institutional Grants for Research Training in Biomedical Informatics	0.2	-
HHS	NIH, OD	Science Education Partnership Award	15.4	-
HHS	NIH, Office of Science Education	Office of Science Education K-12 Program	2.2	-
HHS	HRSA	Public Health Traineeship	1.6	-
HHS	NIH, NIGMS	RISE (Research Initiative for Scientific Enhancement)	6.3	6.2
HHS	NIH	Ruth L. Kirschstein National Research Service Award Institutional Research Training Grants (T32, T35)	487.1	487.1

Appendix Table A6. (Continued)

Agency	Sub-Agency	Investment	FY 12 Enacted	FY 14 Requested
HHS	NIH	Ruth L. Kirschstein NRSA for Individual Predoctoral Fellows, including Underrepresented Racial/Ethnic Groups,Students from Disadvantaged Backgrounds, and Predoctoral Students with Disabilities	4.0	3.7
HHS	NIH, NIDA	Science Education Drug Abuse Partnership Award	3.6	-
HHS	NIH, NICHD	Short Courses in Population Reseach (Education Programs for Population Research R25)	0.1	0.8
HHS	NIH, NIGMS	Short Courses on Mathematical, Statistical, and Computational Tools for Studying Biological Systems	0.5	-
HHS	NIH, NIEHS	Short Term Educational Experiences for Research (STEER) in the Environmental health Sciences for Undergraduates and High School Students	0.4	-
HHS	NIH, NHLBI	Short-Term Research Education Program to Increase Diversity in Health-Related Research	2.9	0.1
HHS	NIH, Intramural Training	Student Intramural Research Training Award Program	0.2	0.2
HHS	NIH, NHLBI	Summer Institute for Training in Biostatistics	0.1	0.1

Agency	Sub-Agency	Investment	FY 12 Enacted	FY 14 Requested
HHS	NIH, Intramural Training	Undergraduate Scholarship Program for Individuals from Disadvantaged Backgrounds	0.1	0.1
HHS Total			**575.7**	**530.8**
Homeland Security	S&T Office of University Programs	Homeland Security STEM Career Development Grant Program	2.7	-
Homeland Security	DNDO	National Nuclear Forensics Expertise Development Program	5.5	5.6
Homeland Security	S&T Office of University Programs	Scientific Leadership Awards Program	2.9	3.0
Homeland Security Total			**11.1**	**8.6**
Interior	Bureau of Land Management	Conservation and Land Management Internship Program	1.6	1.5
Interior	USGS	Educational Component of the National Geologic Mapping Program (EDMAP)	0.6	0.6
Interior	National Park Service	George Melendez Wright Climate Change Youth Initiative	0.4	0.4
Interior	National Park Service	Geoscientists-in-the-Parks Program	0.3	0.2

Appendix Table A6. (Continued)

Agency	Sub-Agency	Investment	FY 12 Enacted	FY 14 Requested
Interior Total			**3.0**	**2.7**
NASA	ARMD	Aeronautics Academy	0.5	-
NASA	ARMD	Aeronautics Content - Smart Skies/Product Content Upgrade	0.8	-
NASA	ARMD	Aeronautics Scholarship	1.8	-
NASA	Education Office	AESP - Aerospace Education Services Project	3.1	-
NASA	Science Mission Directorate (SMD)	Aqua	0.3	-
NASA	Science Mission Directorate (SMD)	Astrophysics Forum	1.0	-
NASA	Science Mission Directorate (SMD)	Aura	0.3	-
NASA	Science Mission Directorate (SMD)	Cassini	0.9	-
NASA	Center JSC	CEP - Career Exploration Project	1.1	-

Agency	Sub-Agency	Investment	FY 12 Enacted	FY 14 Requested
NASA	Science Mission Directorate (SMD)	Chandra	0.5	-
NASA	Center JPL	Curriculum Improvement Partnership Award for the Integration of Research into the Undergraduate Curriculum (CIPAIR)	1.6	-
NASA	Science Mission Directorate (SMD)	DAWN	0.3	-
NASA	ARMD	Design Competitions	0.1	-
NASA	Science Mission Directorate (SMD)	Earth Science E/PO Forum	0.9	-
NASA	Center MSFC	eEducation Small Projects/Central Operation of Resources for Educators (CORE)	0.7	-
NASA	Center JSC	EFP - Education Flight Projects	2.2	-
NASA	Science Mission Directorate (SMD)	EPOESS	6.6	-
NASA	ESMD	ESMD Space Grant Project	0.6	-

Appendix Table A6. (Continued)

Agency	Sub-Agency	Investment	FY 12 Enacted	FY 14 Requested
NASA	Education Office	GCCE - Global Climate Change Education	3.2	-
NASA	Science Mission Directorate (SMD)	GLOBE Program	4.5	4.5
NASA	Science Mission Directorate (SMD)	GRAIL	0.4	-
NASA	Education Office	GSRP - Graduate Student Researchers Program	2.8	-
NASA	Science Mission Directorate (SMD)	Heliophysics E/PO Forum	0.7	-
NASA	SOMD	HEOMD-Goldstone Apple Valley Radio Telescope (GAVRT) Project	0.5	-
NASA	ESMD	HEOMD-NASA's Beginning Engineering, Science and Technology (BEST) Students (NBS)	0.4	-
NASA	ESMD	HEOMD-University Student Launch Initiative	0.3	-
NASA	Science Mission Directorate (SMD)	HST	1.6	-

Agency	Sub-Agency	Investment	FY 12 Enacted	FY 14 Requested
NASA	Education Office	Informal STEM Education	10.0	-
NASA	ARMD	Innovation in Aeronautics Instruction Competition	0.1	-
NASA	Education Office	Innovation in Higher Education STEM Education	0.5	-
NASA	Education Office	INSPIRE - Interdisciplinary National Science Program Incorporating Research and Education Experience	0.7	-
NASA	Education Office	JPFP - Jenkins Pre-Doctoral Fellowship Program	2.6	-
NASA	Science Mission Directorate (SMD)	Juno	0.6	-
NASA	Center LaRC	LARSS - NASA Langley Aerospace Research Summer Scholars Program	0.6	-
NASA	Science Mission Directorate (SMD)	LDCM	0.6	-
NASA	Education Office	LEARN - Learning Environment and Research Network	3.0	-

Appendix Table A6. (Continued)

Agency	Sub-Agency	Investment	FY 12 Enacted	FY 14 Requested
NASA	Center GRC	LERCIP - Lewis Educational Research Collaborative Internship Project (College)	0.1	-
NASA	Education Office	LTP - Learning Technologies Project	0.5	-
NASA	Science Mission Directorate (SMD)	Mars E/PO Formal Ed	1.1	-
NASA	Science Mission Directorate (SMD)	Mars E/PO Informal Ed	1.0	-
NASA	Science Mission Directorate (SMD)	MESSENGER	0.3	-
NASA	Education Office	MUREP (4 STEM programs in FY 2012 Inventory: MUREP Small Projects, NASA Science and Technology Institute for Minority Institutions, NASA Tribal College and University Project, University Research Centers)	18.8	30.0
NASA	Center GRC	MUST - Motivating Undergraduates in Science and Technology	2.3	-
NASA	Center JSC	NAS - NASA Aerospace Scholars	0.3	-
NASA	Education Office	NES - NASA Explorer Schools	3.8	-

Agency	Sub-Agency	Investment	FY 12 Enacted	FY 14 Requested
NASA	Center MSFC	NETS - NASA Education Technologies Services	1.0	-
NASA	ESMD	NSBRI Higher Education Activities - National Space Biomedical Research Institute	0.8	-
NASA	Science Mission Directorate (SMD)	Planetary Science E/PO Forum	0.9	-
NASA	SOMD	Reduced Gravity Student Flight Opportunity Project	0.3	-
NASA	Education Office	Research Cluster	1.4	-
NASA	ESMD	SEED - Systems Engineering Educational Discovery	0.3	-
NASA	Center GRC	SEMAA - Science Engineering Mathematics and Aerospace Academy/FIRST Buckeye	1.0	-
NASA	Science Mission Directorate (SMD)	SOFIA (Stratospheric Observatory for Infrared Astronomy) Education and Public Outreach	0.6	-
NASA	Education Office	SOI - Summer of Innovation/NASA IV&V Engineering Apprenticeship Program	5.9	-

Appendix Table A6. (Continued)

Agency	Sub-Agency	Investment	FY 12 Enacted	FY 14 Requested
NASA	Education Office	Space Grant - National Space Grant College and Fellowship Program	40.0	24.0
NASA	OCT-ST	Space Technology Research Fellowships	12.0	15.0
NASA	CMO/ARC	Spaceward Bound	0.4	-
NASA	Education Office	STEM Accountability and Coordination (formerly STEM Education & Accountability Project)	-	26.7
NASA	Education Office	USRP - Undergraduate Student Research Project	0.3	-
NASA Total			**149.4**	**100.2**
NRC	Office of the Chief Human Capital Officer	Integrated University Program	15.0	-
NRC	Small Business and Civil Rights Office	Minority Serving Institutions Program (MSIP)	0.7	0.7
NRC	Office of the Chief Human Capital Officer	Grants to Universities (Curriculum Development) Program	-	-
NRC Total			**15.7**	**0.7**

Agency	Sub-Agency	Investment	FY 12 Enacted	FY 14 Requested
NSF	Directorate for Education and Human Resources (EHR)	Advanced Informal STEM Learning (AISL), formerly Informal Science Education (ISE)	61.4	47.8
NSF	Directorate for Education and Human Resources (EHR)	Advanced Technological Education (ATE)	64.0	64.0
NSF	Directorate for Education and Human Resources (EHR)	Alliances for Graduate Education and the Professoriate (AGEP)	7.8	7.8
NSF	Directorate for Education and Human Resources (EHR)	Catalyzing Advances in Undergraduate STEM Education (CAUSE)	-	123.1
NSF	Directorate for Geosciences (GEO)	Centers for Ocean Sciences Education Excellence	4.2	0.9
NSF	Directorate for Education and Human Resources (EHR)	Climate Change Education (CCE)	10.0	-
NSF	Directorate for Computer & Information Science & Engineering (CISE)	Computing Education for the 21st Century (CE21)	15.0	-

Appendix Table A6. (Continued)

Agency	Sub-Agency	Investment	FY 12 Enacted	FY 14 Requested
NSF	Office of Cyberinfrastructure (OCI)	Cyberinfrastructure Training, Education, Advancement, and Mentoring for Our 21st Century Workforce (CI-TEAM)	4.0	-
NSF	Directorate for Education and Human Resources (EHR)	Discovery Research K-12 (DR-K12)	99.2	102.5
NSF	Office of International & Integrative Activities (OIIA)	East Asia & Pacific Summer Institutes for U.S. Graduate Students (EAPSI)	2.4	2.4
NSF	Directorate for Engineering (ENG)	Engineering Education (EE)	11.1	-
NSF	Directorate for Math and Physical Sciences (MPS)	Enhancing the Mathematical Sciences Workforce in the 21st Century (EMSW21)	11.8	11.5
NSF	Directorate for Education and Human Resources (EHR)	Excellence Awards in Science and Engineering (EASE)	5.2	4.8
NSF	Directorate for Education and Human Resources (EHR)	Federal Cyber Service: Scholarship for Service (SFS)	45.0	25.0

Agency	Sub-Agency	Investment	FY 12 Enacted	FY 14 Requested
NSF	Directorate for Geosciences (GEO)	Geoscience Education	1.5	-
NSF	Directorate for Geosciences (GEO)	Geoscience Teacher Training (GEO-Teach)	2.0	-
NSF	Directorate for Geosciences (GEO)	Global Learning and Observations to Benefit the Environment (GLOBE)	1.1	-
NSF	Directorate for Education and Human Resources (EHR) & Office of International and Integrative Activities (OIIA)	Graduate Research Fellowship Program (GRFP)	198.1	325.1
NSF	Directorate for Education and Human Resources (EHR)	Graduate Teaching Fellows in K-12 Education (GK-12)	27.0	-
NSF	Directorate for Education and Human Resources (EHR)	Historically Black Colleges and Universities Undergraduate Program (HBCU-UP)	31.9	31.9
NSF	Directorate for Education and Human Resources (EHR)	Innovative Technology Experiences for Students and Teachers (ITEST)	25.0	25.0

Appendix Table A6. (Continued)

Agency	Sub-Agency	Investment	FY 12 Enacted	FY 14 Requested
NSF	Directorate for Education and Human Resources (EHR)	Integrative Graduate Education and Research Traineeship (IGERT) Program	59.8	-
NSF	Office of International & Integrative Activities (OIIA)	International Research Experiences for Students (IRES)	3.2	2.3
NSF	Directorate for Education and Human Resources (EHR)	Louis Stokes Alliances for Minority Participation (LSAMP)	45.6	45.6
NSF	Directorate for Education and Human Resources (EHR)	Math and Science Partnership (MSP)	57.1	-
NSF	Directorate for Engineering (ENG)	Nanotechnology Undergraduate Education in Engineering	1.5	-
NSF	NSF	NSF Research Traineeships (NRT)	-	55.1
NSF	Directorate for Education and Human Resources (EHR)	NSF Scholarships in Science, Technology, Engineering, and Mathematics (S-STEM)	75.0	75.0

Agency	Sub-Agency	Investment	FY 12 Enacted	FY 14 Requested
NSF	Directorate for Geosciences (GEO)	Opportunities for Enhancing Diversity in the Geosciences	3.6	-
NSF	Directorate for Engineering (ENG)	Research Experiences for Teachers (RET) in Engineering and Computer Science	5.5	5.5
NSF	Directorate for Education and Human Resources (EHR)	Research Experiences for Undergraduates (REU)	66.0	79.2
NSF	Directorate for Education and Human Resources (EHR)	Research in Disabilities Education (RDE)	6.5	-
NSF	Directorate for Education and Human Resources (EHR)	Research on Education and Learning (REAL), formerly Research and Evaluation on Education in Science and Engineering (REESE)	37.7	60.4
NSF	Directorate for Education and Human Resources (EHR)	Research on Gender in Science and Engineering (GSE)	10.5	-
NSF	Directorate for Education and Human Resources (EHR)	Robert Noyce Scholarship (Noyce) Program	54.9	60.9

Appendix Table A6. (Continued)

Agency	Sub-Agency	Investment	FY 12 Enacted	FY 14 Requested
NSF	Directorate for Education and Human Resources (EHR)	Science, Technology, Engineering, and Mathematics Talent Expansion Program (STEP)	25.3	-
NSF		STEM-C Partnerships	-	73.6
NSF	Directorate for Education and Human Resources (EHR)	Transforming Undergrad Education in STEM (TUES)	39.5	-
NSF	Directorate for Biological Sciences (BIO)	Transforming Undergraduate Biology Education (TUBE)	13.0	-
NSF	Directorate for Education and Human Resources (EHR)	Tribal Colleges and Universities Program (TCUP)	13.3	13.3
NSF	Directorate for Education and Human Resources (EHR)	Widening Implementation and Demonstration of Evidence-based Reforms (WIDER)	8.0	-
NSF Total			**1,153.7**	**1,242.8**
Smithsonian		STEM Informal Education and Instruction	-	25.0

Agency	Sub-Agency	Investment	FY 12 Enacted	FY 14 Requested
Smithsonian Total			-	25.0
Transportation	Federal Aviation Administration (FAA)	Air Transportation Centers of Excellence	12.5	12.5
Transportation	Research and Innovative Technology Administration (RITA)	Dwight David Eisenhower Transportation Fellowship Program	1.9	1.5
Transportation	Research and Innovative Technology Administration (RITA)	Garrett A. Morgan Technology and Transportation Education Program	1.1	0.9
Transportation	Federal Highway Administration (FHWA) - Office of Civil Rights	National Summer Transportation Institute Program (STI)	2.6	3.0
Transportation	Research and Innovative Technology Administration (RITA)	Summer Transportation Institute Program for Diverse Groups (STIPDG)	1.3	1.5
Transportation	Research and Innovative Technology Administration (RITA)	University Transportation Centers Program	80.0	72.5
Transportation Total			99.4	91.9
Federal Total			2,890.7	3,069.0

APPENDIX B

Investment Design Principles

Applying these principles in the design and implementation of investments is intended to improve the prospects of accomplishing the investment's primary objective and enhance the potential for stronger cross-agency coordination. These principles represent what might be considered useful frameworks for to allow more Federal investments to be successful, based on currently available evidence and best practice. Agencies will be at a variety of stages in their capacity and readiness to enact these principles, and as agencies work to do so, new principles will emerge and the current set will be refined.

There are three categories of design principles:

- general investment design principles that may apply to all investments
- design principles by objective for each of the main CoSTEM investment objective categories
- design principles for investments serving underrepresented groups

Federal STEM education investments should consider applying *general investment design principles* and the appropriate *design principles by objective.* Investments that aim to support groups traditionally underrepresented in STEM education should look to incorporate the *design principles for investments serving underrepresented groups*.

Over the period of implementation of the Strategic Plan, agency staff should collaborate with other agencies or non-Federal STEM education experts about how to incorporate design principles into their planning for existing and new investments, and how to build the expertise and capacity to do so over time. The activity of conforming to design principles will evolve as programs build that expertise and capacity. Thus, as implementation of the Strategic Plan begins investments will be assessed based on their progress in increasing capacity to align with the design principles.

General Investment Design Principles

CoSTEM recommends that agencies create or regularly update logic models or theory-of-action documents, management plans, and evaluation strategies for their investments, and address how their investments incorporate the applicable design principles.

As appropriate, investments should have a logic model or explicit theory of action that describes:

- Clear overarching goals, specific investment objectives and measurable outcomes
- How the investment helps to fulfill the agency's mission and connects to agency education or STEM assets[145], and STEM education goals
- Results of needs assessments, stakeholder input, or environmental scans that helped shape the goals, objectives, and outcomes
- Alignment with evidence-based practices, promising practices of experienced professionals, and relevant education research

- Strategic partnering within the agency, or with other Federal agencies, education organizations, or stakeholders, or why partnerships are not appropriate
- How the investment activities are likely to advance the goals and objectives

As appropriate, investments should have a management plan that describes:

- Needed expertise, staffing plan and how expertise is assured, plans for professional development of staff to implement design principles, and assignment of accountability for outcomes
- Strategies for dissemination or scale-up of promising practices and lessons learned in the course of implementation
- Plans for budget allocation and cost-sharing
- Strategic partnering within the agency, or with other Federal agencies, education organizations, or stakeholders, or why partnerships are not appropriate
- Resources needed for evaluation and staffing

As appropriate, investments should have an evaluation strategy that describes:

- A prioritized list of evaluation questions and goals
- Alignment of evaluation with the agency's STEM education evaluation strategy or agency strategic performance plan
- Evaluation methods, design and measures that are appropriate for the questions to be addressed, the outcomes being measured, the type of investment, its stage of implementation, and other relevant factors
- Data to be gathered and methods of data collection
- Use of standards for program evaluation[146] or evaluation research design
- Plans for sharing results of evaluation and identifying promising approaches
- Formal mechanisms by which evaluation findings inform improvements in investments

STEM Education for Underrepresented Groups Principles

In order to increase the number of individuals from underrepresented groups that graduate with STEM degrees, STEM investments should:

- Be designed with input from and implemented with the participation of underrepresented groups, local community stakeholders, and other stakeholders as appropriate
- Draw upon, relate to, and be respectful of the interests, knowledge, practices, and culturally relevant STEM experiences of underrepresented groups and demonstrate an understanding of targeted communities based on research
- Take advantage of place-based or experiential learning opportunities and agency STEM professionals, facilities, technology, and data where appropriate, particularly for mission agencies
- Build sustained relationships between participants and STEM partners (including continuous tracking and mentoring of participants)

- Build capacity and develop a strategy or mechanism to build a robust STEM workforce pipeline

Design Principles by Primary Investment Objective

The Federal STEM Education Portfolio Review (NSTC, 2011) requests that agencies identify a "single primary objective" for each investment (p. 13). In addition to applying the General Investment Design Principles, and those for underrepresented groups if applicable, investments should also consider applying the design principles listed here according to primary objective.

Learning Investments and Engagement Investments

Learning Investments "develop STEM skills, practices, or knowledge of students or the public." (NSTC, 2011, p. 13) Engagement Investments are designed to increase learners' involvement and interest in STEM, inform their view of the value of STEM in their lives, or enhance their ability to participate in STEM. These investments will address learners both inside and outside of formal education systems, at all levels. A single set of design principles can apply to both types of investments. Learning or Engagement Investments should work to:

- Have activities articulate learning goals describing what students or the public should know and be able to do as a result of engaging with the education activity, as well as goals for building interest, developing identity, improving motivation, and other affective outcomes, as appropriate
- Build from and acknowledge the significant expertise and professional knowledge of designers and developers of learning materials and activities
- Incorporate expectations for pilot, laboratory, or field testing of materials and activities, and iterative improvement and redesign
- Acknowledge the policy and learning contexts in which materials and activities would be used, with attention to such matters as national, state, or local standards; local or state education agency criteria; and available scientific expertise or facilities
- Encourage use of evidence-based practices and draw on effective approaches (e.g., active learning, career assessment, mentoring, coaching, cohort or team experiences, peer learning, authentic science experiences, citizen science, experiential, hands-on, inquiry, or place-based learning)
- Expect that learning and affective outcomes will be assessed, if possible using common approaches or instruments, across activities in the investment
- Include, as appropriate, comprehensive and coherent support or interface materials (e.g., teacher manuals, aligned assessments, affiliated websites, leadership supports, online platforms, etc.)

Pre- and in-Service Educator and Leader Performance

Investments in this area are designed to "train or retain STEM educators (K-12 pre-service or in-service, postsecondary, and informal) and education leaders to improve their content knowledge and pedagogical skills." (NSTC, 2011, p. 13)

Investments in pre-service educator or leader preparation should:

- Anticipate changing contexts for STEM educators or leaders (e.g., student demographics, learning technologies, accessible new scientific discoveries and innovations)
- Acknowledge the appropriate policy context for educator or leader preparation (e.g., college- and career-readiness standards, teacher preparation accreditation, teacher evaluation, and state licensure and credentialing processes)
- Reflect understanding of research developments about educator preparation, including the role of STEM disciplinary knowledge, STEM knowledge for teaching, effective classroom practices in STEM education, clinical experiences and practices, and assessment of STEM learning

In addition, investments in in-service educator and leader development should:

- Expect that activities are connected to systems of in-service educator or leader support (e.g. instructional materials, workshops, assessments, coaching, administrative expertise in STEM education)
- Assess training needs for activity staff who interact with learners and provide professional development
- For educators, align with specific local or state education agency professional development systems or priorities or provide professional credentials
- Serve multiple teachers or leaders in a school

Postsecondary STEM Degrees, Workforce Development, and STEM Careers

Postsecondary STEM degree investments are aimed at increasing the number of students who enroll in STEM majors, complete STEM credential or degree programs, or are prepared to enter STEM careers or advanced education. *STEM career investments* comprise general and agency mission-specific workforce education programs to prepare people to enter the STEM workforce with training or certification. STEMdiscipline-specific knowledge and skills are the primary focus of the education investment.

Investments in either category should:

- Address specific STEM workforce capacity and expertise needs based on evidence from labor market data or workforce planning information
- Build involvement of representatives from potential employment and workforce sectors in design
- Establish strategic recruitment approaches that reach diverse audiences

- Have projects include appropriate mentoring, educational and internship experiences, academic support, and follow-up opportunities for networking and professional connections
- Incorporate authentic agency or STEM industry work or related experience

In addition, scholarship or fellowship investments should:

- Demonstrate that financial incentives such as scholarships or fellowships provide value added in attracting and retaining students to the workforce, if applicable
- Ensure sufficient support so that students successfully complete STEM programs or degrees

Institutional Capacity

Several Federal STEM education investments are designed to support the advancement and development of STEM personnel, programs, and infrastructure. This occurs in universities, informal education institutions, state education agencies, and local education agencies.

Investments should:

- Identify the institutional accountability, policy and incentive systems that may need transformation to support the goal, and recognize the importance of institutional context
- Justify choice of target institutions and provide rationale for why investment in these institutions is likely to leverage broader impact
- Build demonstration of commitment by institutional leadership
- Clarify "change strategies" for making improvements
- Have an analysis of resource needs at the institutional level
- Create clear sustainability strategies to extend beyond funding period

Education Research and Development

Investments in this area support the development of evidence-based STEM education models and practices. Such investments should:

- Incorporate explicit expectations about such key elements of research as theoretical and empirical bias, study design, data sources, and plans for interpretation and dissemination
- Encourage applicants to employ shared evidence standards, including efforts by NSF and IES

Include mechanisms for gathering feedback from small-scale implementers and practitioners that informs more basic R&D, to ensure that findings from basic research inform program development and implementation, and to help with relevance and usability.

End Notes

[1] http://www.whitehouse.gov/sites/default/files/docs/stem_teachers_release_3-18-13_doc.pdf

[2] Pub. L. No. 111-358 (http://www.gpo.gov/fdsys/pkg/BILLS-111hr5116enr/pdf/BILLS-111hr5116enr.pdf)

[3] Pub. L. No. 111-358 (http://www.gpo.gov/fdsys/pkg/BILLS-111hr5116enr/pdf/BILLS-111hr5116enr.pdf).

[4] Members of the FC-STEM are listed on p. 6.

[5] Members of CoSTEM are listed on pages 6, 7. The same agencies were represented on the FI-STEM.

[6] NSTC (2011). The Federal STEM Education Portfolio. http://www.whitehouse.gov/sites/default/files/microsites/ostp/costem__Federal_stem_education_portfolio_report.pdf.

[7] STEM fields are defined in the National Science Foundation's Science and Engineering Indicators, http://www.nsf.gov/statistics

[8] PCAST President's Council of Advisors on Science and Technology. (February 2012). Report to the President: Engage to excel: Producing one-million additional college graduates with degrees in science, technology, engineering, and mathematics. http://www.whitehouse.gov/sites/default/files/microsites/ostp/pcast-engage-to-excelfinal_2-25-12.pdf

[9] Fleischman, H.L., Hopstock, P.J., Pelczar, M.P., and Shelley, B.E. 2010. *Highlights From PISA 2009: Performance of U.S. 15-Year-Old Students in Reading, Mathematics, and Science Literacy in an International Context* (NCES 2011-004). U.S. Department of Education, National Center for Education Statistics.

[10] Committee on Underrepresented Groups and the Expansion of the Science and Engineering Workforce Pipeline. 2010. *Expanding Underrepresented Minority Participation: America's Science and Technology Talent at the Crossroads.*Committee on Science, Engineering, and Public Policy; Policy and Global Affairs; National Academy of Sciences, National Academy of Engineering, and Institute of Medicine. Available from http://www.nap.edu/catalog/12984.html.

[11] Economics and Statistics Administration. (2011). Women in STEM: A Gender Gap to Innovation. United States Department of Commerce, Washington, D.C.

[12] This fifth priority, focused on the STEM workforce, was discussed in CoSTEM and was not specified as a goal in the progress report of February 2012. It has been recast as a goal in light of the President's FY 2014 STEM education reorganization. It has since been added in the development of this final report.

[13] Economics and Statistics Administration. (2011). STEM: Good Jobs Now and for the Future. United States Department of Commerce, Washington, D.C.

[14] U.S. Department of Commerce (January, 2012). The competitiveness and innovative capacity of the United States. http://www.commerce.gov/sites/default/files/documents/2012/january/competes_010511_0.pdf.

[15] http://www9.georgetown.edu/grad/gppi/hpi/cew/pdfs/stem-complete.pdf

[16] Business-Higher Education Forum. (2007). An American Imperative Transforming the Recruitment, Retention, and Renewal of Our Nation's Mathematics and Science Teaching Workforce. Retrieved from http://www.eric.ed.gov/PDFS/ED503709.pdf.

[17] Tai, R. H., Liu, Q. C., Maltese, A. V., & Fan, X. (2006). Planning early for careers in science. *Science*. 312, 1143 - 1144.

[18] Martin, M.O., Mullis, I.V.S., Foy, P., & Stanco, G.M. (2012). Chestnut Hill, MA: TIMSS & PIRLS International Study Center, Boston College.

[19] Carnevale, A.P., Smith, Nicole, and Melton, M. *STEM*. 2011. Georgetown University, Washington, D.C. http://www9.georgetown.edu/grad/gppi/hpi/cew/pdfs/STEMWEBINAR.pdf

[20] PCAST President's Council of Advisors on Science and Technology. (February 2012). Report to the President: Engage to excel: Producing one-million additional college graduates with degrees in science, technology, engineering, and mathematics. http://www.whitehouse.gov/sites/default/files/microsites/ostp/pcast-engage-to-excelfinal_2-25-12.pdf

[21] http://www.nsf.gov/statistics/seind12/pdf/c02.pdf

[22] National Research Council and National Academy of Engineering. *Community Colleges in the Evolving STEM Education Landscape: Summary of a Summit*. 2012. Washington, DC: The National Academies Press.

[23] National Center for Education Statistics. (2011). (Table Illustration Digest of Education Statistics May 3, 2013). *Number of persons age 18 and over, by highest level of educational attainment, sex, race/ethnicity, and age:2011.* Retrieved from http://nces.ed.gov/programs/digest/d11/tables/dt11_009.asp.

[24] http://nces.ed.gov/pubs2011/2011317.pdf

[25] Division of Science Resources Statistics. *Women, Minorities, and Persons with Disabilities in Science and Engineering: 2011.* Special Report NSF 11-309. Arlington, VA.

[26] http://www.whitehouse.gov/sites/default/files/microsites/ostp/pcast-engage-to-excel-final_2-25-12.pdf

[27] Federal Coordinating Council for Science, Engineering, and Technology, Committee on Education and Human Resources (1993). Pathways to excellence: A Federal strategy for science, mathematics, engineering, and technology education. http://www.eric.ed.gov/PDFS/ED360165.pdf.

[28] NSTC Subcommittee on Education (2008). Finding out what works: Agency efforts to strengthen the evaluation of Federal STEM education programs. http://www.whitehouse.gov/files/documents/ostp/NSTCpercent 20Reports/ NSTC_Education_Report_Complete.pdf .

[29] National Research Council. (2008). *NASA's Elementary and Secondary Education Program: Review and Critique.* Committee for the Review and Evaluation of NASA's Precollege Education Program, Helen R. Quinn, Heidi A. Schweingruber and Michael A. Feder, Editors. Board on Science Education. Washington, DC: The National Academies Press.

[30] National Research Council. (2010). *NOAA's Education Program: Review and Critique.* Committee for the Review of the NOAA Education Program. J.W. Farrington and M.A. Feder, Editors. Board on Science Education. Washington, DC: The National Academies Press.

[31] http://www2.ed.gov/programs/racetothetop/index.html

[32] http://www2.ed.gov/programs/innovation/index.html

[33] Clotfelter, C.T., Ladd, H.F., and Vigdor, J.L. (2007). Teacher credentials and student achievement in high school: A cross-subject analysis with student fixed effects. *Economics of Education Review, 26*(6), 673-782.

[34] Rivkin, S.G. (2007). *Value added analysis and education policy.* Washington, DC: Urban Institute, Center for the Analysis of Longitudinal Data in Education Research.

[35] Boyd, D., Grossman, P., Lankford, H., Loeb, S., Rockoff, J., and Wyckoff, J. (2008). The narrowing gap in New York City: Teacher qualifications and its implications for student achievement in high-poverty schools. *Journal of Policy Analysis and Management, 27*(4), 793-818.

[36] Rockoff, J.E. (2004). The impact of individual teachers on student achievement: Evidence from panel data. *American Economic Review, 94*(2), 247-252.

[37] Hanushek, E. A. & Rivkin, S.G. The Distribution of Teacher Quality and Implications for Policy. *Annual Review of Economics,* 4, September 2012, pp. 131-157.

[38] Hanushek, Eric A. The Economic Value of Higher Teacher Quality. *Economics of Education Review,* 30(3), June 2011, pp. 466-479.

[39] Ingersoll, R., & Perda, D. (2010a). How high is teacher turnover and is it a problem? Philadelphia: University of Pennsylvania, Consortium for Policy Research in Education.

[40] Ingersoll, R.M. and Perda, D. (2010b). Is the supply of Mathematics and Science Teachers Sufficient? *American Educational Research Journal*, 47(3), 563-594.

[41] Tai, R. H., Liu, Q. C., Maltese, A. V., & Fan, X. (2006). Planning early for careers in science. Science. 312, 1143 - 1144.

[42] Bybee, R. W., Taylor, J., Gardner, A., Scotter, P., Powell, J., Westbrook, A. et al. (2006). *The BSCS 5E Instructional Model: Origins, Effectiveness, and Application.* Colorado Springs, CO: Commissioned Paper, National Institutes of Health, Office of Science Education.

[43] *Many Experts, Many Audiences: Public Engagement with Science and Informal Science Education*, with Ellen McCallie, Larry Bell, Tiffany Lohwater, John H. Falk, Brice V. Lewenstein, Cynthia Needham, and Ben Wiehe, A CAISE Inquiry Group Report, 2009.

[44] National Research Council and National Academy of Engineering, *Community Colleges in the Evolving STEM Landscape: Summary of a Summit*, 2012.

[45] Carnevale, A.P., N.Smith, and J. Strohl. (2010). Help Wanted: Projections of Jobs and Education Requirements through 2018. Washington, DC: Georgetown University Center on Education and the Workforce.

[46] Lacey, T. A. and B. Wright. (2009). "Occupational employment projections to 2018." Monthly Labor Review 132(11):82-123.

[47] Langdon, D., G. McKittrick, D. Beede, B. Khan, and M. Doms. (2011). "STEM: Good Jobs Now and for the Future." ESA Issue Brief #03-11. Washington, DC: U.S. Department of Commerce.

[48] U. S. Census Bureau (2011a), Population Division, "Table 3. Annual Estimates of the Resident Population by Sex, Race, and Hispanic Origin for the United States: April 1, 2000 to July 1, 2009 (NC-EST2009-03)."

[49] U.S. Census Bureau (2011b), Population Estimates,http://www.census.gov/popest .

[50] U.S. Census Bureau (2009). U.S. Population Projections. Retrieved May 24, 2012 from http://www.census.gov/population/projections.html .U.S. Census, 2009

[51] http://pathwaysreport.org/rsc/pdf/ex_summary.pdf

[52] http://acd.od.nih.gov/biomedical_research_wgreport.pdf, http://www.fgereport.org/rsc/pdf/CFGE_report.pdf, http://www.nap.edu/rdp/ http://portal.acs.org/portal/PublicWebSite/about/governance/CNBP_031603

[53] This model, proposed in the President's FY 2013 Budget and included in the Senate appropriations bill for Labor/HHS/Education for programs serving disconnected youth, allows multiple Federal programs to blend funds to support outcome-focused strategies and to waive statutory requirements that are not necessary to meet outcome goals or ensure appropriate use of funds.

[54] Clotfelter, C.T., Ladd, H.F., and Vigdor, J.L. (2007). Teacher credentials and student achievement in high school: A cross-subject analysis with student fixed effects. *Economics of Education Review, 26*(6), 673-782.

[55] Rivkin, S.G. (2007). *Value added analysis and education policy.* Washington, DC: Urban Institute, Center for the Analysis of Longitudinal Data in Education Research.

[56] Boyd, D., Grossman, P., Lankford, H., Loeb, S., Rockoff, J., and Wyckoff, J. (2008). The narrowing gap in New York City: Teacher qualifications and its implications for student achievement in high-poverty schools. *Journal of Policy Analysis and Management, 27*(4), 793-818.

[57] Rockoff, J.E. (2004). The impact of individual teachers on student achievement: Evidence from panel data. *American Economic Review, 94*(2), 247-252.

[58] National Center for Education Statistics. Accessed on January 3, 2013 at http://nces.ed.gov/programs /projections/projections2020/tables/table_16.asp

[59] (Council of Chief State School Officers, Washington, DC, 2007)

[60] National Center for Alternative Certification, Accessed at http://www.teach-now.org/intro.cfm on April 22, 2013

[61] Ingersoll, R. and Perda, D. (2010). Is the Supply of Mathematics and Science Teachers Sufficient?, *American Educational Research Journal*, Vol. 43(3). pp. 563-594.

[62] Hess, F., Kelly, A., and Meeks, O. (2011). The Case for Being Bold A New Agenda for Business in Improving STEM Education. http://www.aei.org/papers/education/the-case-for-being-bold/ (April 2011).

[63] Schmidt et al.(2010). Breaking the Cycle: An International Comparison of U.S. Mathematics Teacher Preparation, Michigan State University Center for Research in Mathematics and Science Education.

[64] Ball, D., Hill, H., and Bass, H. (2005). Knowing Mathematics for Teaching. *American Educator*. Fall 2005. pp. 14-22, 43-46.

[65] National Research Council. *Preparing Teachers: Building Evidence for Sound Policy*. Washington, DC: The National Academies Press, 2010.

[66] http://www.whitehouse.gov/sites/default/files/microsites/ostp/pcast-stemed-report.pdf, 59.

[67] Ball, D., Hill, H., and Bass, H. (2005). Knowing Mathematics for Teaching. *American Educator*. Fall 2005. pp. 14-22, 43-46.

[68] National Science and Technology Council, (December 2011). *The Federal Science, Technology, Engineering, and Mathematics (STEM) Education Portfolio. Washington, DC.*

[69] National Research Council. (2011). *STEM Smart Brief STEM Smart: Lessons Learned From Successful Schools.* Committee on Highly Successful Science Programs for K-12 Science Education. Board on Science Education and Board on Testing and Assessment, Division of Behavioral and Social Sciences and Education. Washington, DC: The National Academies Press.

[70] Ingersoll, R. and Perda, D. (2010). Is the Supply of Mathematics and Science Teachers Sufficient?, *American Educational Research Journal*, Vol. 43(3). pp. 563-594.

[71] Authentic STEM experiences may be provided to K-12 teachers via internships, fellowships, and scholarships; "grade-appropriate" acknowledges that the nature of authentic STEM experiences is likely to differ for PK-5, 6-8 and 9-12 grade educators.

[72] Authentic STEM experiences may be provided to P-12 teachers via internships, fellowships, and scholarships; "grade-appropriate" acknowledges that the nature of authentic STEM experiences is likely to differ for PK-5, 6-8 and 9-12 grade educators.

[73] L. Deslauriers, E. Schelew, C. Wieman, *Science* 332**,** 862 (2011).

[74] B. A. Nagda, S. R. Gregerman, J. Jonides, W. von Hippel, J. S. Lerner, *Rev. Higher Educ.* 22, 55 (1998).

[75] See, for example, Sivan, Leung, Woon & Kember (2000), An implementation of active learning and its effect on the quality of student learning. *Innovations in Education and Training International 37* (4): 381-389.

[76] Might the NRC Learning Science in Informal Settings report

[77] Knox, K. L., Moynihan, J. A., & Markowitz, D. G. (2003). Evaluation of Short-Term Impact of a High School Summer Science Program on Students' Perceived Knowledge and Skills. *Journal of Science Education & Technology, 12*(4), 471-478; Markowitz, D. G. (2004). Evaluation of the Long-Term Impact of a University High School Summer Science Program on Students' Interest and Perceived Abilities in Science. *Journal of Science Education & Technology, 13*(3), 395-407.

[78] http://www.nasa.gov/offices/education/programs/national/dln/special/DigitalBadges.html

[79] http://www.nationalservice.gov/newsroom/press-releases/2013/president-obama-announces-stem-americorpsinspire-young-peoples

[80] http://www.mnh.si.edu/education/yes/about.html

[81] http://us2020.org/

[82] Committee on STEM Education, *The Federal Science, Technology, Engineering and Mathematics Education Portfolio,* Washington, D.C. National Science Technology Council.

[83] Efforts should reflect the recent OSTP memorandum on Increasing Access to the Results of Federally Funded Scientific Research http://www.whitehouse.gov/sites/default/files/microsites/ostp/ostp_public_access_memo_2013.pdf

[84] Carnevale, A.P., N.Smith, and J. Strohl. (2010). Help Wanted: Projections of Jobs and Education Requirements through 2018. Washington, DC: Georgetown University Center on Education and the Workforce.

[85] Lacey, T. A. and B. Wright. (2009). "Occupational employment projections to 2018." Monthly Labor Review 132(11):82-123.

[86] Langdon, D., G. McKittrick, D. Beede, B. Khan, and M. Doms. (2011). "STEM: Good Jobs Now and for the Future." ESA Issue Brief #03-11. Washington, DC: U.S. Department of Commerce.

[87] http://my-goals.performance.gov/sites/default/files/images/STEM%20Education%20CAP%20Goal%20- %20FY 2012%20Quarter%204%20Update_2.pdf

[88] National Research Council. (2012). *Education for Life and Work: Developing Transferable Knowledge and Skills in the 21st Century.* Committee on Defining Deeper Learning and 21st Century Skills

[89] National Research Council and National Academy of Engineering, *Community Colleges in the Evolving STEM Landscape: Summary of a Summit*, 2012.

[90] National Research Council, *Discipline-Based Education Research: Understanding and Improving Learning in Undergraduate Science and Engineering*, 2012.

[91] PCAST (February 2012). Report to the President: Engage to excel: Producing one-million additional college graduates with degrees in science, technology, engineering, and mathematics. http://www.whitehouse.gov /sites/default/files/microsites/ostp/pcast-engage-to-excel-final_2-25-12.pdf

[92] http://cep.mit.edu

[93] http://my-goals.performance.gov/sites/default/files/images/STEM%20Education%20CAP%20Goal%20- %20FY 2012%20Quarter%204%20Update_2.pdf

[94] http://www.nsf.gov/news/news_summ.jsp?cntn_id=127902&org=NSF&from=news

[95] Langdon, D.. McKittrick, G. Beede, D. Khan, B., Julian, T., Lehrman, R., and Doms, M. (2011). Education Supports Racial and Ethnic Equality in STEM. ESA Issue Brief #05-11. Washington, DC: U.S. Department of Commerce.

[96] Langdon, D., McKittrick, G., Beede, D., Khan, B., and Doms, M. (2011). Women in STEM: A Gender Gap to Innovation. ESA Issue Brief #04-11. Washington, DC: U.S. Department of Commerce.

[97] C. Loes ***et al.***, Effects of diversity experiences on critical thinking skills over four years of college; www.education.uiowa.edu/centers/docs/cdre-documents/Loes_Pascarella_and_Umbach_2012_3.pdf?sfvrsn=0

[98] S. E. Page, The Difference: How the Power of Diversity Creates Better Groups, Firms, Schools, and Societies (Princeton Univ. Press, Princeton, NJ, 2007).

[99] U.S. Department of Education, National Center for Education Statistics, Integrated Postsecondary Education Data System (IPEDS), Fall 2010, Completions component (prepared November 2011), *available at* http://nces.ed.gov/programs/digest/d11/tables/dt11 301.asp.

[100] *ibid.*

[101] *ibid.*

[102] Wei, et al., "Science, Technology, Engineering, and Mathematics (STEM) Participation Among College Students with an Autism Spectrum Disorder," Journal of Autism and Developmental Disorders, November 1, 2012, Table 4, *available at* http://link.springer.com/article/10.1007/s10803-012-1700-z/fulltext.html

[103] The data from the CRDC, while covering 85% of the nation's public school students, are not intended to be an estimation of national data. All data in the CRDC are self-reported.

[104] Alvarado, C. & Dodds, Z Women in CS: an evaluation of three promising practices, Proceedings of the 41st ACM technical symposium on Computer science education, 2010

[105] Walton, G. M. & Spencer, S. J. (2009). Latent ability: Grades and test scores systematically underestimate the intellectual ability of negatively stereotyped students. *Psychological Science, 20,* 1132-1139.

[106] Walton, G. M., & Cohen, G. L. (2011). A brief social-belonging intervention improves academic and health outcomes of minority students. *Science,331*(6023), 1447-1451.

[107] *Employment Projections: 2010-2020*, Bureau of Labor Statistics, 2012.

[108] PCAST (February 2012). Report to the President: Engage to excel: Producing one-million additional college graduates with degrees in science, technology, engineering, and mathematics. http://www.whitehouse.gov /sites/default/files/microsites/ostp/pcast-engage-to-excel-final 2-25-12.pdf

[109] *Survey of Doctorate Recipients*, NSF/National Center for Science and Engineering Statistics, 2008.

[110] *Science and Engineering Indicators 2012* Arlington, VA (NSB 12-01) | January 2012

[111] http://grants.nih.gov/grants/guide/pa-files/PA-11-184.html

[112] Investing in the Future: NIGMS Strategic Plan for Biomedical and Behavioral Research Training; Blueprint for Implementation. http://www.nigms.nih.gov/Training/StrategicPlanImplementationBlueprint.htm

[113] National Research Council. "Goals for U.S. STEM Education." *Successful K-12 STEM Education: Identifying Effective Approaches in Science, Technology, Engineering, and Mathematics*. Washington, DC: The National Academies Press, 2011.

[114] National Research Council. (2012). *Discipline-Based Education Research: Understanding and Improving In Undergraduate Science and Engineering*. Committee on the Status, Contributions, and Future Directions of Discipline-Based Education Research. Board on Science Education Washington, DC: The National Academies Press.

[115] http://ies.ed.gov/ncee/projects/evaluation/tq.asp

[116] Charles T. Clotfelter, Helen F. Ladd, and Jacob L. Vigdor (2013), *Algebra for 8th Graders: Evidence on its Effects from 10 North Carolina Districts*, working paper

[117] http://grants.nih.gov/training/career_progress/index.htm

[118] National Research Council. (2007). *Taking Science to School: Learning and Teaching Science in Grades K-8*. Washington, DC: The National Academies Press.

[119] National Research Council. (2009). *Learning Science in Informal Environments: People, Places, and Pursuits*. Committee on Learning Science in Informal Environments. Philip Bell, Bruce Lewenstein, Andrew W. Shouse, and Michael A. Feder, Editors. Board on Science Education, Washington, DC: The National Academies Press.

[120] National Research Council. (2011). *Successful K-12 STEM Education: Identifying Effective Approaches in Science, technology, Engineering, and Mathematics*. Committee on Highly Successful Science Programs for K-12 Science Education. Board on Science Education and Board on Testing and Assessment. Washington, DC: The National Academies Press.

[121] National Research Council. (2011). *Successful K-12 STEM Education: Identifying Effective Approaches in Science, technology, Engineering, and Mathematics*. Committee on Highly Successful Science Programs for K-12 Science Education. Board on Science Education and Board on Testing and Assessment. Washington, DC: The National Academies Press.

[122] National Research Council. (2012). *Discipline-Based Education Research: Understanding and Improving In Undergraduate Science and Engineering*. Committee on the Status, Contributions, and Future Directions of Discipline-Based Education Research. Board on Science Education Washington, DC: The National Academies Press.

[123] National Research Council (2001). *Expanding Underrepresented Minority Participation: America's Science and Technology Talent at the Crossroads*. Committee on Science, Engineering, and Public Policy. Washington, DC: National Academy Press.

[124] National Mathematics Advisory Panel. (2008). Foundations for Success The Final Report of the National Mathematics Advisory Panel. Retrieved from http://www2.ed.gov/about/bdscomm/list/mathpanel/report/final-report.pdf.

[125] National Research Council. (2010). *Preparing teachers: Building evidence for sound policy.* Committee on the Study of Teacher Preparation Programs in the United States. Division of Behavioral and Social Sciences and Education. Washington, DC: The National Academies Press.

[126] President's Council of Advisors on Science and Technology. (September, 2010). Prepare and inspire: K-12 education in science, technology, engineering, and mathematics (STEM) for America's future. http://www.whitehouse.gov/sites/default/files/microsites/ostp/pcast-stemed-report.pdf.

[127] President's Council of Advisors on Science and Technology (February 2012). Report to the President: Engage to excel: Producing one-million additional college graduates with degrees in science, technology, engineering, and mathematics. http://www.whitehouse.gov/sites/default/files/microsites/ostp/pcast-engage-to-excel-final_2-25-12.pdf

[128] National Research Council. (2008). *NASA's Elementary and Secondary Education Program: Review and Critique*. Committee for the Review and Evaluation of NASA's Precollege Education Program, Helen R. Quinn, Heidi A.

[129] National Research Council. (2010). *NOAA's Education Program: Review and Critique*. Committee for the Review of the NOAA Education Program. J.W. Farrington and M.A. Feder, Editors. Board on Science Education. Washington, DC: The National Academies Press.

[130] Friedman, A. (Ed.). (March 12, 2008). Framework for Evaluating Impacts of Informal Science Education Projects [On-line]. Retrieved from http://insci.org/resources/Eval_Framework.pdf.

[131] http://findyouthinfo.gov/docs/Common%20Standards-Draft%2002-28-13_508_3-11-13_Author.pdf

[132] http://www.regulations.gov/#!documentDetail;D=NSF-2012-OTR-0002-0001

[133] http://www.whitehouse.gov/sites/default/files/omb/memoranda/2012/m-12-14.pdf

[134] Analytic support was provided by the Science & Technology Policy Institute, which is a federally funded research and development center (FFRDC) chartered by Congress (https://www.ida.org/stpi.php)

[135] http://www.whitehouse.gov/sites/default/files/microsites/ostp/costem__federal_stem_education_portfolio_ report.pdf

[136] http://cahsi.cs.utep.edu/Portals/0/Resources/Literature/Report_of_the_AcademicCompetitiveness_Council.pdf

[137] http://www.afa.org/ProfessionalDevelopment/AerospaceEducation/ReportsandStudies/RS_PDFs/InventorySurvey.pdf

[138] FY 2008-2011 data were collected

[139] The total funding amount in figures is noted only if it differs from the FY 2011 sum of $2,891 million.

[140] While only one primary objective could be selected, multiple secondary objectives could be selected.

[141] The categories in this table are not mutually exclusive and thus cannot be summed.

[142] "Adults" does not include investments for educators

[143] Partnership and Interagency Collaboration information was only collected for broader STEM Education investments.

[144] More than one evaluation design could be identified for each investment.

[145] STEM and education expertise, resources, networks, facilities, policies

[146] See, for example, American Evaluation Association program evaluation standards, http://www.eval.org

INDEX

D

M

N

O

P

Q

R

S

T